Making Sense of People

Also by Samuel Barondes

Cellular Dynamics of the Neuron

Neuronal Recognition

Molecules and Mental Illness

Mood Genes

Better Than Prozac

Before I Sleep: Poems for Children Who Think

Making Sense of People

The Science of Personality Differences

Samuel Barondes

Second Edition

Publisher: Paul Boger
Editor-in-Chief: Amy Neidlinger
Executive Editor: Jeanne Glasser Levine
Cover Designer: Alan Clements
Managing Editor: Kristy Hart
Project Editor: Andy Beaster
Copy Editor: Language Logistics, LLC
Proofreader: Chuck Hutchinson
Indexer: Joy Lee
Compositor: Nonie Ratcliff
Manufacturing Buyer: Dan Uhrig

Published by Pearson Education, Inc.
Old Tappan, New Jersey 07675

For information about buying this title in bulk quantities, or for special sales opportunities (which may include electronic versions; custom cover designs; and content particular to your business, training goals, marketing focus, or branding interests), please contact our corporate sales department at corpsales@pearsoned.com or (800) 382-3419.

For government sales inquiries, please contact governmentsales@pearsoned.com.

For questions about sales outside the U.S., please contact international@pearsoned.com.

Printed in the United States of America

First Printing November 2015

ISBN-10: 0-13-421500-1
ISBN-13: 978-0-13-421500-6

Pearson Education LTD.
Pearson Education Australia PTY, Limited
Pearson Education Singapore, Pte. Ltd.
Pearson Education Asia, Ltd.
Pearson Education Canada, Ltd.
Pearson Educación de Mexico, S.A. de C.V.
Pearson Education—Japan
Pearson Education Malaysia, Pte. Ltd.

Library of Congress Control Number: 2015948629

For Louann

And for my grandchildren:
Jonah Lazar
Ellen Ariel
Asher Lucca

Contents

Every man is in certain respects

(a) like all other men,

(b) like some other men,

(c) like no other man.

—Clyde Kluckhohn and Henry A. Murray

Preface to Second Edition

In 2001, after meeting Vladimir Putin for the first time, George W. Bush offered his famous impression of the Russian's personality:[1]

> I looked the man in the eye. I found him to be very straightforward and trustworthy and we had a good dialogue. I was able to get a sense of his soul.

Bush's assessment of Putin as straightforward and trustworthy triggered various interpretations. Some took it at face value and were relieved that the two world leaders had hit it off. Others felt that Bush had been deceived by the ex-KGB man and were alarmed by his gullibility. Still others just dismissed it as the sort of polite statement politicians make to warm up their adversaries.

Years later Bush made clear he had meant what he said. When questioned directly by Hugh Hewitt during the 2010 book tour for his memoir, *Decision Points*, Bush explained it this way:[2]

> The reason why I said that is because I remembered him talking movingly about his mother and the cross she gave him that she had blessed in Jerusalem. Nobody knows that, and I never tried to make an explanation of why I said what I said until the book.

But Bush also saw a different side of Putin, which he revealed in a portrait he made after retiring from the presidency.[3] In contrast with his earlier assessment, the portrait does not depict Putin as being particularly trustworthy or straightforward. Instead the man Bush painted has been called poker-faced and unreadable, scary and sinister, brimming with anger, contempt, and disgust.

How did Bush arrive at this darker view of the Russian leader? Was it based on his rethinking of Putin's conduct on the world stage?

Although that would seem likely, the answer Bush gave was more personal: It had to do with Putin's disdain for Bush's beloved dog Barney. On two occasions Putin had offended Bush by mocking Barney's weakness and small size and by comparing him unfavorably with his own dog, Koni. In response Bush apparently painted the face of Putin as the face of the man who had ridiculed his dear little friend. As Bush explained it in 2014 on the *NBC Today Show*:[4]

> Wow, anybody who thinks "My dog is bigger than
> your dog" is an interesting character. And the paint-
> ing kind of reflects that.

The painting, of course, reflects more than "an interesting character." The painting also reflects the difference between what Bush is comfortable expressing on canvas and what he is comfortable expressing in words.

George W. Bush is not alone. Many of us find it difficult to articulate our views of personalities—not only to others, but also to ourselves. There's so much to consider, and it's hard to convert what we know in our minds into a useful verbal picture.

Nevertheless, we can learn to do better. Making well-crafted portraits with words is just as teachable as making them with oils or pastels. In each case what's needed is good instruction and a dedicated student.

In the first edition of this book, I explained a step-by-step system for making better verbal portraits. It combined four ways of thinking about personalities based on decades of research by psychologists and psychiatrists. It showed how the information from these four perspectives could be put together into a rich and complex picture of each unique person.

Many readers found this helpful, but others had difficulty integrating the various parts. They wanted more practical assistance in applying it to the people in their lives. Put simply, they wanted more "how to."

This new edition is designed with those people in mind. The only substantial change I've made is to add a section, called "Practical Summary," at the end of each chapter to discuss and illustrate applications. In these sections I also address misunderstandings and controversial points. The result I've aimed for is not only more "how to." It's also more "here's why." At the end of the last chapter, I sum up the benefits of integrating information from all four perspectives into an overall portrait.

As a further aid to making the book more meaningful, I would like you to pick a significant person in your life (whom I'll call P) and keep him or her in mind as you go through the book. Repeatedly relating the material to this person may help you see what each perspective adds to the whole. To keep reminding you to do this, I will ask you to answer some relevant questions about P at the end of each Practical Summary.

Getting Started with P

To prepare for this focus on P, here are the steps to take now:

1. Pick a person you've spent a lot of time with, preferably someone who is at least 25 years old. Make this choice carefully because I'd like you to stick with it until the end of the book.

2. Think back on the first time you met P and on important experiences you've shared. Mull this over and notice what characteristics of P come to mind.

3. Now write a description of P's personality using words, phrases, sentences, and full paragraphs as you see fit. Record the description on paper or an electronic device and, when you're done, please don't change it. It will serve as a reminder of your starting point to go back to when you've finished the book. But feel free to keep a separate set of notes about P as you go from chapter to chapter.

4. When you're satisfied with your description, which may be as long or short as you like, read on.

When Intuition Isn't Enough

All of us are personality experts. Ever since childhood, we've been paying attention to people's distinctive ways of being and trying to figure out what to expect from them. We depend on this information to get along.

Our innate ability to size people up is an amazing gift we take for granted. With it, we form an instantaneous impression of the personality of everyone we meet. Most of our assessments of people are formed in this automatic and unconscious way.[1]

But there are times when we want to take a closer look by consciously and systematically evaluating someone's personality.[2] We may, for example, want to understand what it is about our boss that makes us avoid her. We may want to sort through the reasons we don't approve of our teenage daughter's boyfriend. We may want to decide if the person we're dating has the right stuff for a permanent relationship.

That's when the going gets tough. The difficulty mainly arises because few of us have been taught a systematic way to assess personalities. Instead, we are constantly bombarded with a contradictory mishmash of religious, moral, literary, and psychological ideas that are hard to apply in an orderly manner. Imagine how we would struggle to do simple arithmetic if we kept getting contradictory instructions on how to work with numbers. Yet we're expected to make sense of people without having been taught a coherent arithmetic of personality.

This lack of education may be responsible for some of our biggest mistakes. It can lead us to pick the wrong suitor, take the wrong job, or misguide our children. It can cause us to misinterpret a coworker's intentions and become inappropriately defensive, or compliant, or aggressive. It can keep us from building satisfying relationships, gracefully avoiding conflicts, or developing plans to protect our interests by fighting back.

In this book, I describe a system for thinking about personalities that may help you avoid such mistakes. Based on decades of research, each chapter will make it easier for you to organize the data you already have about particular people and to start noticing characteristics that you may have overlooked. Sorting through this information will give you a clearer sense of each person and how to relate to them.

To get started, I will show you how to combine two vocabularies that professionals use to organize their observations. One breaks down personality into five well-defined tendencies, such as conscientiousness and agreeableness, each of which has several components. This makes it easier to think things through using a well-defined set of words.

The other vocabulary shifts attention from these tendencies to ten potentially troublesome patterns of behavior, such as compulsiveness or paranoia. Mild versions of these patterns may simply be notable parts of a well-functioning personality. But some of us have inflexible and maladaptive versions of one or a few of them, versions that frequently bring grief to those we deal with—and to ourselves. More than the rest of us, such people are prisoners of personality who are locked into ways of being they seem unable to escape.

Combining these two easy-to-learn vocabularies will not only help you make clearer assessments of everyone you meet, but will also raise questions about the reasons people get to be so different from each other. In the second part of the book, I will describe the development of the brain circuits that control our distinctive combinations of tendencies and patterns. I will also show that the decades-long developmental process that builds these brain circuits is strongly influenced by the two great accidents of our birth: the specific set of genes we happen to be born with and the specific world we happen to live in.

But there's more to a personality than tendencies and patterns. In the third part of the book, I will turn to the values and goals that give meaning and purpose to people's lives. To flesh out this view, I will show you how to apply universal and culture-specific standards of morality to assess that aspect of personality called character. I will also encourage you to pay attention to the stories people tell about their personal history and future plans, which will help you figure out what they stand for and their sense of identity.

Systematically organizing all this information about tendencies, patterns, character, and identity will help you make sense of anyone. It may also influence the approach you choose to engage with them. In some cases, this may encourage you to shrug off their disquieting idiosyncrasies in favor of forgiveness and compassion. In other cases, it may alert you to telltale signs of danger so that you can take protective actions. In still other cases, it may open your heart to warm feelings of love and respect. In all cases, it will enhance your appreciation of human diversity in the same way that those who know a lot about wine, or music, or baseball get the added pleasure that comes from thoughtful attention to the details. Augmenting your pleasure in understanding and dealing with people, whether you like them or not, is the main aim of this book.

PART I

Describing Personality Differences

The beginning of wisdom is to call things by their right names.

—Chinese Proverb

ONE

Personality Traits

When I was in high school, I signed up for the student newspaper. To get me started, the editor offered some standard advice on how to write a story. He said I should be sure to answer five questions: What happened? Who was involved? When? Where? Why? He said that knowing about these "five Ws" served as a check for completeness because novices sometimes left out one or more of them. He then assured me that I wouldn't need them for long because answering these questions was something I was already inclined to do intuitively.

Intuition is also what journalists rely on when they size up people. Through years of practice, they develop a knack for identifying distinctive personality traits and finding the words to describe them. The gifted among them are so good at it that they can create a revealing portrait in a single paragraph. Consider, for example, Joe Klein's description of the personality of an American politician:

> There was a physical, almost carnal, quality to his
> public appearances. He embraced audiences and was
> aroused by them in turn. His sonar was remarkable in
> retail political situations. He seemed able to sense
> what audiences needed and deliver it to

them—trimming his pitch here, emphasizing different priorities there, always aiming to please. This was one of his most effective, and maddening qualities in private meetings as well: He always grabbed on to some point of agreement, while steering the conversation away from larger points of disagreement—leaving his seducee with the distinct impression that they were in total harmony about everything. ...
There was a needy, high cholesterol quality to it all; the public seemed enthralled by his vast, messy humanity. Try as he might to keep in shape, jogging for miles with his pale thighs jiggling, he still tended to a raw fleshiness. He was famously addicted to junk food. He had a reputation as a womanizer. All of these were of a piece.[1]

Notice that Klein needs only a handful of evocative words to highlight the main characteristics of his subject: carnal, needy, messy, maddening, fleshiness, addicted, and womanizer. To round out his description, he uses a few short phrases, such as "his sonar was remarkable," "high cholesterol quality," and "aiming to please." When he can't find a simple word or phrase to describe something that he considers particularly revealing, he makes up a whole sentence: "he always grabbed on to some point of agreement, while steering the conversation away from larger points of disagreement—leaving his seducee with the distinct impression that they were in total harmony about everything." By using words and phrases that all of us can understand, Klein tells us a great deal about the personality of an extraordinary public figure: Bill Clinton.

The combination of words and phrases is, of course, critical. There are other people who are needy but who are neither carnal nor womanizers. Some of them may also have remarkable sonar but without being messy or maddening. What makes Klein's description so recognizable is that, as he points out, all the traits "were of a piece."

So how did Klein do it? Was he intuitively asking himself a set of questions that are as obvious to him as the five Ws? Did he leave out anything important? Can we learn a technique to make our own descriptions of people more incisive and complete?

Words from the Dictionary

The development of a simple technique to describe personalities was set in motion in the 1930s by Gordon Allport, a professor of psychology at Harvard. Although Allport was well aware of the uniqueness of each individual, he also knew that scientific fields get started by breaking down complex systems into simple components. Just as understanding the great variety of chemical compounds depended on identifying a limited number of elements, understanding the great variety of personalities may depend on identifying a limited number of critical ingredients. But what exactly are those ingredients?

Allport's answer was traits: the enduring dispositions to act and think and feel in certain ways that are described by words found in all human languages. Just as chemical elements such as carbon and hydrogen can combine with many others to form endless numbers of complicated substances, traits such as being outgoing and being reliable can combine with many others to form endless numbers of complicated personalities. But how many traits are there? And how could Allport find out?

To answer this question, Allport and his colleague, H.S. Odbert, made a list of the words about personality from *Webster's New International Dictionary*.[2] By analyzing this list, they hoped to identify the essential components of personality that were so obvious to our ancestors that they invented a great many words to describe them. Instead of just concocting an inventory of personality traits out of their own heads, Allport and Odbert would be guided by the cumulative verbal creations of countless minds over countless generations, as recorded in a dictionary.[3]

It soon became clear that these researchers had bitten off more than they could chew. The list of words "to distinguish the behavior of one human being from another" had 17,953 entries! Faced with this staggering number, they whittled it down using several criteria. First, they eliminated about a third, such as *attractive,* because the entries were considered evaluative rather than essential: "[W]hen we say a woman is attractive, we are talking not about a disposition 'inside the skin' but about her effect on other people."[4] Another fourth hit the cutting room floor because they described temporary states of mind, such as *frantic* and *rejoicing,* rather than the enduring dispositions that are defining features of personality traits. Others were thrown out because they were considered ambiguous. In the end, about 4,500 entries met the researchers' criteria for stable traits.

This doesn't mean that personality has 4,500 different components; many of the words on the list are easily identifiable as synonyms. For example, *outgoing* and *sociable* are used interchangeably. Furthermore, antonyms, such as *solitary*, describe the same general category of behavior, but at its opposite pole—instead of saying "not sociable" or "not outgoing," we might say "solitary." In fact, a wonderful feature of natural language is that it lends itself so well to a graded (or dimensional) description of specific components of personality, from extremely outgoing at one pole to extremely solitary at the other, with modifiers to specify points in between. Put simply, the ancestors who gradually built our language—and all languages—left us with many choices for describing ingredients of personality.

Recognizing that *outgoing* and *solitary* both refer to aspects of an identical trait, how many other words also fit into this category? When I looked up *outgoing* in my thesaurus, I found these synonyms, among others: *gregarious, companionable, convivial, friendly,* and *jovial.* When I looked up *solitary,* I got, among others, *retiring, isolated, lonely, private,* and *friendless.* This tells me that the group of experts who put together this thesaurus decided that all these words belong in a box that can be labeled Outgoing–Solitary. Needless to say, each word in the box may also have some special spin of its own. For example, *solitary, lonely,* and *private* don't mean exactly the same thing, and writers such as Joe Klein may mull them over to get just the right one. Nevertheless, we all know that these words have a lot in common. To psychologists such as Allport, they all refer to a single overarching trait.

Beyond Synonyms and Antonyms

Does this mean that we can identify the essential building blocks of personality by simply getting a list from a dictionary and then lumping together the synonyms and antonyms from a thesaurus? Can we base a nomenclature of personality on the analysis of professional lexicographers? Or can we use a more open-source approach that pays attention to the ways ordinary people employ words to describe personalities?

The answer psychologists settled on was both. First, professionals reduced the list to a more manageable number—about a thousand. Then they asked ordinary people to use these words to describe themselves and their acquaintances. To get an idea of the way this was done, please apply the ten words in the following list to someone you know well. In expressing your opinion, use a scale of 1 to 7, with 7 indicating that the person ranks very high, 1 indicating that the person ranks very low, and the other numbers indicating that the person falls somewhere in between.

1. Outgoing   1 2 3 4 5 6 7
2. Bold   1 2 3 4 5 6 7
3. Talkative   1 2 3 4 5 6 7
4. Energetic   1 2 3 4 5 6 7
5. Assertive   1 2 3 4 5 6 7
6. Reliable   1 2 3 4 5 6 7
7. Practical   1 2 3 4 5 6 7
8. Hardworking   1 2 3 4 5 6 7
9. Organized   1 2 3 4 5 6 7
10. Careful   1 2 3 4 5 6 7

I have no way of knowing what numbers you selected. But chances are good that they will have a characteristic relationship: The numbers you picked for the first five items probably are similar, and the numbers you picked for the second five items probably are similar. Furthermore, I can say with confidence that most people who give someone a certain score for *outgoing* give them a similar score for *bold, talkative, energetic,* and *assertive;* and that the score they give someone for *reliable* is likely similar to the one they give for *practical, hardworking, organized,* and *careful.* Even though none of the words in each quintet are synonyms, people who are ranked a certain way on one word from each tend to get similar scores on the others. In contrast, people's scores on the first quintet are independent of their scores on the second quintet. This implies that these nonsynonymous words are grouped together in our minds because each refers to some aspect of a related component of personality.

Could any other words be lumped together with *outgoing* or *reliable* to flesh out these two big categories? How many other groupings like this would be discovered if people were asked to make judgments using all the thousand words that the original list was pared down to? And what statistical techniques would be needed to identify these categories? In making the list, Allport set the stage for research on these questions.[5]

Bundling Traits

A statistical technique for studying the relationships between these words was invented in the nineteenth century by Francis Galton, a founder of modern research on personality, whom you read more about later. The technique is used to calculate a correlation coefficient, a number between 1.0 and −1.0 that measures the degree of sameness (positive correlation) or oppositeness (negative correlation). Although Galton invented the technique for other purposes, he also happened to be interested in categorizing the words that we use for personality traits,[6] and he would have been pleased to learn about this application.

To get a feel for this calculation, let's think about the positive correlations we would find if we asked people to rank someone on *outgoing, sociable,* and *gregarious* by using a scale of 1 to 7. Knowing that these words are synonyms, we would expect to find that if John ranks Mary a 6 on *outgoing,* he likely will rank her around 6 on each of the others. If he then ranks Jane as a 4 on *outgoing,* he likely will rank her around 4 on each of the others. And if Jennifer ranks Jim a 1 on *outgoing,* she likely will rank him around 1 on each of the others. Plugging these scores into Galton's formula would indicate a great deal of sameness.

Now what sort of correlations would we find between the words in the first nonsynonymous quintet (outgoing–bold–talkative–energetic–assertive)? Studies show that these words are correlated strongly, but not as strongly as synonyms, and similar positive correlations are found among the words in the second nonsynonymous quintet (reliable–practical–hardworking–organized–careful). In contrast, when we compare the scores for words such as *outgoing* from the first group with words such as *reliable* from the second group, we don't find a correlation. This comes as no surprise because we all know that being outgoing and being reliable are not intrinsically related.

Determining the correlations among five or ten words is fairly easy. But determining the correlations among a thousand words was stalled until researchers could turn it over to a computer. To get the raw data, thousands of ordinary people were asked to apply each of these words by ranking their applicability to themselves or another person using a scale of 1 to 7. The mass of data was then analyzed with a more advanced statistical technique, called *factor analysis*, which measures the correlation between each word and all the others and organizes the correlations into clusters. In this way, some words were identified as highly correlated to each other, making them good representatives of a particular cluster of traits, each of which can be thought of as a general tendency.

By the early 1980s, the results were in: The words that describe personality traits can be boiled down to just five tendencies (see Table 1.1), which Lewis Goldberg named the Big Five.[7] Each of these tendencies has been given a reasonably descriptive name: Extraversion (E), Agreeableness (A), Conscientiousness (C), Neuroticism (N), and Openness (O). If you have trouble recalling these names at first, as I did, you can use the acronyms OCEAN or CANOE to jog your memory until they become second nature.

TABLE 1.1   The Big Five: Representative Words

	High	Low
Extraversion vs. Introversion	Outgoing, bold, talkative, energetic, assertive	Withdrawn, timid, silent, reserved, shy
Agreeableness vs. Antagonism	Warm, kind, cooperative, trusting, generous	Cold, unkind, uncooperative, suspicious, stingy
Conscientiousness vs. Disinhibition	Reliable, practical, hardworking, organized, careful	Unreliable, impractical, lazy, disorganized, negligent
Neuroticism vs. Emotional Stability	Tense, unstable, discontented, irritable, insecure	Relaxed, stable, contented, imperturbable, secure
Openness vs. Closedness	Imaginative, curious, reflective, creative, sophisticated	Unimaginative, uninquisitive, unreflective, uncreative, unsophisticated

Using the Big Five

After the Big Five were discovered, they became the foundation for assessing individual differences in the ways people interact with their social and physical worlds. Three tendencies—Extraversion, Agreeableness, and Neuroticism—mainly relate to ways of interacting with other people. The other two—Conscientiousness and Openness—are more general.[8]

- **Extraversion is the tendency to actively reach out to others.** People high in Extraversion are stimulated by the social world, like to be the center of attention, and often take charge. They also like excitement and are inclined to be upbeat, fun loving, full of energy, and to experience positive emotions. People low in Extraversion are less interested in interpersonal interactions and tend to be reserved and quiet. But their relative lack of interest in being with people need not indicate that they don't like them or that they are socially anxious or depressed; they may just prefer to be alone.

- **Agreeableness is the tendency to be altruistic, cooperative, and good-natured.** People high in Agreeableness are considerate, compassionate, helpful, and willing to compromise. They truly like people and assume that everyone is decent and trustworthy. People low on Agreeableness are more self-interested than altruistic, more competitive than cooperative, and likely to be skeptical of others' intentions. They also tend to be cold, antagonistic, and disrespectful of the rights of others.

- **Conscientiousness is the tendency to control impulses and to tenaciously pursue goals.** People high in Conscientiousness are orderly, reliable, hardworking, neat, and punctual. They tend to plan ahead and think things through. They are more interested in long-term than short-term goals. People low in Conscientiousness are more spontaneous, less constrained, less dutiful, and less achievement-oriented. Although Conscientiousness shows up prominently in the performance of tasks, it also influences interpersonal relationships.

- **Neuroticism is the tendency to have negative feelings, particularly in reaction to perceived social threats.** People high in Neuroticism are emotionally unstable, tend to be upset by minor threats or frustrations, and are often in a bad mood. They are prone to anxiety, depression, embarrassment, self-doubt, self-consciousness, anger, and guilt. People low on Neuroticism are emotionally stable, calm, composed, and unflappable. But their freedom from negative feelings does not imply that they are particularly inclined to have positive feelings.

- **Openness is the tendency to be imaginative and to enjoy novelty and variety.** People who are high in Openness tend to be artistic, nonconforming, intellectual, aware of their feelings, and comfortable with new ideas. People low in Openness prefer the simple, straightforward, familiar, and obvious to the complex, ambiguous, novel, and subtle. They tend to be conventional, conservative, and resistant to change. Although people who are high on Openness enjoy the life of the mind, Openness is not identical to intelligence. Highly intelligent people can be high or low on O.

After you've mulled over the broad meanings of these five domains, you can get a better sense of them by applying them to someone you know. You might start by asking yourself how outgoing, good-natured, reliable, moody, and creative that person is compared with others. In doing this, you will notice that the person's relative rankings vary somewhat depending on the situation.[9] For example, a person may be outgoing with friends but shy with strangers, so you have to decide on the average scores by summing up the many observations you've made.[10] From this, you will come away with a profile of the person's basic tendencies, such as moderately extraverted, very agreeable and conscientious, a little neurotic, and very open. Although this is no more than a rough summary of how you regard this person, the Big Five framework will have helped you put your intuitive assessments into words. You will then be in a position to more thoughtfully compare this person with others by seeing his or her differences more clearly.[11]

Big Five 2.0

Having made such assessments, you may find that your ideas about each category are still fuzzy. To sharpen your appraisal of a person's profile of traits, it helps to move from a holistic impression to a more meticulous examination. To do this, you need to learn more about the details of the Big Five.

Paul Costa and Robert McCrae have done the most to clarify these details. Working together at the National Institutes of Health in the 1980s, they developed a questionnaire called the NEO PI-R, which uses phrases rather than adjectives.[12] The big advantage of using phrases is that you can design them to eliminate some of the ambiguity that is inherent in single words. For example, in place of the word *insecure,* a component of Neuroticism, Costa and McCrae use phrases that spell out certain aspects, such as "In dealing with people, I always dread making a social blunder" and "I often feel helpless and want someone else to solve my problems."[13]

Another reason for the popularity of the NEO PI-R is that it sharpens the assessment of each of the Big Five by subdividing them into six components, called facets. This ensures a more complete evaluation and helps focus attention on specific individual differences. Consider, for example, these phrases that assess facets of Extraversion:

- I find it easy to smile and be outgoing with strangers. (Warmth/Friendliness)

- I enjoy parties with lots of people. (Gregariousness)

- I am dominant, forceful, and assertive. (Assertiveness)

- My life is fast-paced. (Activity)

- I love the excitement of roller coasters. (Excitement-Seeking)

- I am a cheerful, high-spirited person. (Positive Emotions/Cheerfulness)

The advantage of using these facets is that it may help you make distinctions that you might have glossed over. For example, many people with an average E score are not average across the board. Some may be somewhat higher on warmth, gregariousness, and positive emotions than on assertiveness, activity, and excitement-seeking; others may have a different balance. The same is true for the other tendencies. In each case, you should pay particular attention to facets that stand out as clearly higher or lower than average. Because the whole point of the exercise is to compare people with each other, you're really looking for these distinguishing characteristics. You may also take note of particular situations in which these distinguishing characteristics are expressed.

To get a feel for the facets of the Big Five, I encourage you to take a free computer-based personality test that resembles the proprietary one devised by Costa and McCrea, at www.personal.psu.edu/faculty/j/5/j5j/IPIP/ipipneo120.htm. Developed by a group of distinguished personality researchers[14] and overseen by John A. Johnson[15] at Pennsylvania State University, it uses different names for some of the facets but covers similar ground. This free test, called the IPIP, can be taken anonymously in about 20 minutes. If you take it, you will receive an automated email report that shows your relative rankings on the Big Five and its facets by comparing your scores with those of the hundreds of thousands of other people who have already taken it.

To gain more experience with the facets of the Big Five (Table 1.2), you may also use the online questionnaire to assess the familiar person (P) I asked you to select in the Preface. Doing this will sharpen your view of P and help persuade you of the usefulness of this approach. As you become more familiar with the Big Five, the aim is to learn to do some rough scoring in your head (high, medium, or low) without having to rely on a questionnaire.

TABLE 1.2   Facets of the Big Five*

Extraversion

Warmth/Friendliness (makes friends easily)

Gregariousness (likes the company of others)

Assertiveness (likes to take charge)

Activity (likes to be busy)

Excitement-Seeking (likes thrills)

Positive Emotions/Cheerfulness (is prone to feel happy)

Agreeableness

Trust (assumes people have good intentions)

Straightforwardness/Morality (is candid, avoids deception)

Altruism (finds helping others rewarding, is not exploitative)

Compliance/Cooperation (prefers compromise to opposition)

Modesty (is not boastful)

Tender-Mindedness/Sympathy (is kind, compassionate)

Conscientiousness

Competence/Self-Efficacy (can accomplish things)

Order/Orderliness (is well organized, makes plans)

Dutifulness (is highly reliable)

Achievement-Striving (works to achieve excellence)

Self-Discipline (has willpower)

Deliberation/Cautiousness (takes time making decisions)

Neuroticism

Anxiety (is prone to fearfulness)

Angry Hostility (is prone to feel resentful)

Depression (is prone to feel discouraged, pessimistic)

Self-Consciousness (is shy because of fear of rejection)

Impulsiveness/Immoderation (has difficulty resisting urges)

Vulnerability (loses poise under pressure)

Openness

Fantasy/Imagination (tries to create a more interesting world)

Aesthetics/Artistic Interests (loves beauty in art and nature)

Feelings/Emotionality (is aware of own feelings)

Actions/Adventurousness (is eager to try new activities)

Ideas/Intellect (likes to play with ideas)

Values/Liberalism (is ready to challenge convention)

---

* When the facets have different names in the proprietary (NEO PI-R) and nonproprietary (IPIP) tests, I've listed both.

Rethinking Bill Clinton

Another way to get a feel for the Big Five and its facets is to keep it in mind while re-examining the paragraph from Joe Klein's book that I cited at the start of this chapter. Klein tells us much more about Clinton's personality than he packed into this paragraph. But for our purpose, I mainly stick to those 165 words:

> There was a physical, almost carnal, quality to his public appearances. He embraced audiences and was aroused by them in turn. His sonar was remarkable in retail political situations. He seemed able to sense what audiences needed and deliver it to them—trimming his pitch here, emphasizing different priorities there, always aiming to please. This was one of his

most effective, and maddening qualities in private
meetings as well: He always grabbed on to some
point of agreement, while steering the conversation
away from larger points of disagreement—leaving his
seducee with the distinct impression that they were
in total harmony about everything. ... There was a
needy, high cholesterol quality to it all; the public
seemed enthralled by his vast, messy humanity. Try
as he might to keep in shape, jogging for miles with
his pale thighs jiggling, he still tended to a raw fleshiness. He was famously addicted to junk food. He had
a reputation as a womanizer. All of these were of a
piece.

As noted before, Klein built his description by calling
attention to a few key attributes. But now we can translate
the information that Klein provides into the language of the
Big Five. Needless to say, much more is known about Clinton,
and other observers have painted a somewhat different picture
than Klein did.[16] But let's stick with the paragraph and some
other information from his book to illustrate how the Big Five
and its facets can help us organize our thoughts about Clinton's basic tendencies. To do this, I will concentrate on facets
in which his scores are notably high or low.

Starting with Extraversion is particularly fitting when
considering Clinton because he loves to be the center of attention. Klein emphasizes this with evocative terms for his public
appearances, such as "embraced audiences" and "aroused by
them," which translate into very high scores on gregariousness.
Clinton is also obviously high on assertiveness, which led him

to the most powerful leadership roles, and "womanizer" can be considered partly a reflection of high excitement-seeking. From this and everything else Klein tells us, Clinton ranks high on all facets of Extraversion, and his overall score is at the top of the chart.

Klein also gives us some information about Agreeableness, but Clinton's score isn't quite so obvious. From the paragraph, you may first get the impression that he ranks high on A because he is "always aiming to please." But as you read on, you will realize that he's just telling his "seducees" whatever they want to hear. In the course of his book, Klein gives many other examples of Clinton's deceptiveness, which gives him a low score on straightforwardness. Klein also presents evidence that Clinton's womanizing is exploitative, which lowers his score on altruism and sympathy. When taken together, Clinton's Agreeableness, which appears very high on first meeting him, is lower than it seems.

The information we get about Conscientiousness is limited but revealing. The part about jogging indicates an effort at self-discipline. But this impression is tempered by "try as he might to keep in shape," "raw fleshiness," "addicted to junk food," and "womanizer," which are hardly testimony to high C. So even though Klein's paragraph leaves out Clinton's very high achievement-striving, the lower scores for dutifulness, cautiousness, and deliberation that he documents in other parts of the book combine to give a lower than average ranking on Conscientiousness.

Klein's paragraph tells us little about Neuroticism except for a hint about "messy humanity." Other sections of the book tell us that Clinton can get very angry and out of control, but there's no reason to think of him as being especially prone to negative emotions. In fact, he is unusually capable of brushing off criticism that would make most of us crumble, and he can be cool under extreme fire. When taken together, Clinton ranks below average on Neuroticism.

Openness to experience is also not explicitly considered. This omission is not unusual in brief descriptions of people, even though it may turn out to be a distinguishing feature of their personalities. But Klein makes up for this in the rest of the book by providing us with persuasive evidence that Clinton ranks high on most facets of O.

Of course, much about Clinton doesn't show up in this Big Five profile. But to illustrate the usefulness of this way of describing him, let's compare it with a similar assessment of another president, Barack Obama, as a way of thinking about their differences. Like Clinton, we have had many opportunities to see Obama in action. Furthermore, his two autobiographies fill in many blanks.[17]

In making this comparison, Openness doesn't tell us much. Although Clinton and Obama differ in their scores on certain facets, their overall rankings are both high. But their relative scores on Extraversion, Neuroticism, Agreeableness, and Conscientiousness are informative. When taken together, very different profiles emerge.

Extraversion is particularly notable because Obama's overall score is not only lower than Clinton's, but also lower than the scores of most other successful politicians. Although Obama ranks very high on assertiveness and activity, he is not particularly warm or gregarious. Nor does he show much evidence of positive emotion, even when winning a historic election or a Nobel Prize. Klein, who has also written about Obama's personality, offered evidence of his low E from a politician who helped coach Obama for debates during his first presidential campaign: "He is a classic loner...Usually you work hard at prep, and then everyone, including the candidate, kicks back and has a meal together. Obama would go off and eat by himself. He is very self-contained. He is not needy."[18]

This low neediness is another sign of Obama's difference from Clinton: his very low Neuroticism. Whereas Clinton deserves credit for generally controlling resentfulness and discouragement, Obama doesn't seem to feel them at all, even in the face of strong setbacks. In fact, his remarkable emotional stability, which many admire, has also been criticized as Spock-like. Maureen Dowd, another journalist with a gift for describing personalities, called him "President Cool" and "No Drama Obama."[19]

This coolness might also be taken as a sign of low Agreeableness. But Obama clearly ranks high on several of its facets, especially straightforwardness and a preference for cooperation and compromise. Although he does not exude either altruism or tender-mindedness, his behavior suggests that they are at least average. So even though Obama is not especially high on Agreeableness, I consider him to be higher than he might seem.

Obama's high marks on all six facets of Conscientiousness also distinguish him from Clinton. He ranks especially high on deliberation, examining all sides of a problem. As with other personality traits, this can be seen as a mixed blessing, bringing him praise for his thoughtfulness but criticism that he is too professorial and indecisive.

Considering Obama and Clinton in this way shows how the Big Five can help us organize our intuitive observations by making them explicit. Although the profiles that it generates are sketchy, the process focuses our attention on the full range of tendencies, including some that we might otherwise have overlooked. And as you will see, the findings we make in this way provide a framework for describing the personality patterns discussed in the next chapter.

Practical Summary

In this chapter I've shown you how to organize your initial thoughts about a personality by using the Big Five. To supplement what you've learned, I urge you to take the free online test I mentioned earlier if you haven't already done so: www.personal.psu.edu/faculty/j/5/j5j/IPIP/ipip-neo120.htm

The test is a valuable instrument that resembles those professionals use as part of a formal personality assessment. Taking it, and reviewing the results the program sends back, will tell you how your scores compare with those of the hundreds of thousands of other people in the database. It will also help you apply the Big Five to others.

But you can't use a 120-item questionnaire to size up someone in everyday life. Instead you will have to become sufficiently familiar with the Big Five to make an informal survey in your mind.

AN INFORMAL BIG FIVE SURVEY

To make a mental survey, it's easiest to work in an established sequence. I always use the sequence E-A-C-N-O, as in the Bill Clinton example, and suggest you use it too. As with Clinton, I skim through the facets looking for any that are high or low and wind up with a rough sketch. In his case my assessment was: extremely high E, mixed A and C, lower than average N, and high O.

As you gain experience you, too, will be able to quickly sum up your observations in this way. You may also start looking at the Big Five not only as a collection of traits but also as five *general tendencies* that are operating at particular intensities. To give you some practice in making such quick assessments, please think through the many people you know and try to pick out the ones who rank highest or lowest on each of the Big Five. For convenience here, again, is a summary:

- **Extraversion (vs. Introversion):** The tendency to actively reach out to others.

- **Agreeableness (vs. Antagonism):** The tendency to be altruistic, cooperative, and good-natured.

- **Conscientiousness (vs. Disinhibition):** The tendency to control impulses and tenaciously pursue goals.

- **Neuroticism (vs. Emotional Stability)**: The tendency to have negative feelings, particularly in reaction to perceived social threats.

- **Openness (vs. Closedness)**: The tendency to be imaginative and to enjoy novelty and variety.

This exercise doesn't only give you some practice in assessing these tendencies. It also calls your attention to inconsistencies. For example, you may know a person with high E who, paradoxically, avoids large crowds. Though this doesn't seem to fit with his strong tendency to reach out to others, it turns out that strangers make him uneasy and that he has other signs of high N. The result is that his high E, which operates freely in the company of familiar people, is overridden by a situation that dials up his N.

This example shows that applying the Big Five is not as simple as it might seem. Behavior is greatly influenced by situations, and assessing tendencies must take this into account. Furthermore, the five general tendencies are continuously interacting with each other. Their complex interactions, their situational constraints, and the degree of harmony between them are, in fact, major causes of the distinctiveness and diversity of personalities.

P AND THE BIG FIVE

At the end of the Preface, I asked you to pick a person (P) you know well and to write a description of his or her personality (from this point forward, P will be referred to as *female*). If you didn't get around to it then, please do it now, and continue to keep P in mind as you read on.

Now give P a rough score (such as high, medium, or low) on E, A, C, N, and O.

Having recorded these scores, please answer the following:

1. Do you now find it easier to put your thoughts about P into words?
2. Do you now have a clearer impression of some distinctive aspects of her personality?
3. Did you notice anything you left out of your initial description that you now consider a significant omission?
4. Do you believe some inconsistencies you've noticed may be thought of as conflicts of tendencies?

If you're prepared to spend more time on the Big Five, you may wish to go back to the online test, put yourself in P's shoes, and try to answer each question from what you believe to be her point of view. The report that you receive by return email will turn this into a profile and supplement the informal one you've already made. This will tell you if you picked up something from the test that you missed in your informal evaluation.

As you continue through the book, the Big Five will keep coming up, and you will have many opportunities to increase your facility in using it. Now you're ready to learn to combine it with other ways of thinking about personalities.

TWO

Troublesome Patterns

When we talk about people, we don't just use adjectives, such as *dutiful* or *lazy*. We also use nouns, such as *workaholic* or *slacker*. The adjectives are a way of describing traits that someone has. The nouns are a way of describing categories that someone fits into.

Putting people into categories seems very efficient: A single word or phrase appears to offer a big picture of what a person is like. But words such as *workaholic* aren't really labels for a complete personality. For example, *workaholic* means "one who is addicted to work or who voluntarily works excessively hard and unusually long hours," and *slacker* means "a person who shirks work." So instead of describing a whole person, nouns such as *workaholic* or *slacker* are just ways of emphasizing high or low rankings on a single trait—in this case, a facet of Conscientiousness.

We also have nouns for high and low scorers on the rest of the Big Five. For Extraversion, we have *life of the party* on one end and *loner* on the other; for Agreeableness, there's *altruist* and *misanthrope;* for Neuroticism, *whiner* and *cool cat;* for Openness, *innovator* and *traditionalist.* And we use still other nouns for distinctive combinations. For example, *drama queen,* whose definition in my dictionary includes "overreacts to a minor setback" and "thrives on being the center of attraction," combines high N and high E.

The reason words such as *workaholic* and *drama queen* are so popular is that they're not just shorthand ways of summing up some notable rankings. They also carry extra emotional weight because they evoke images that are more vivid than saying "high C" or "high N and high E." It's like the difference between the abstract description of a long tropical fruit that grows in bunches and turns yellow when ripening, and the enticing picture of a banana. Although *workaholic* and *drama queen* are not as clearly defined as *banana*, they each carry a message that immediately grabs your attention and reduces a complex set of characteristics into a simple image.

Recognizing the usefulness of such evocative words, psychiatrists created a vocabulary for the potentially troublesome personality patterns they observed in their practices. In sorting through them, a committee of experts selected ten that seemed particularly important. These patterns, which I call the Top Ten, are summed up in the *American Psychiatric Association's Diagnostic and Statistical Manual* (DSM)[1] along with the following thumbnail sketches:

- **Antisocial**—A pattern of disregard for and violation of the rights of others

- **Avoidant**—A pattern of social inhibition, feelings of inadequacy, and hypersensitivity to negative evaluation

- **Borderline**—A pattern of instability in interpersonal relationships, self-image, and emotions and marked impulsivity

- **Compulsive (obsessive-compulsive[2])**—A pattern of preoccupation with orderliness, perfectionism, and control

- **Dependent**—A pattern of submissive and clinging behavior related to an excessive need to be taken care of

- **Histrionic**—A pattern of excessive emotionality and attention seeking

- **Narcissistic**—A pattern of grandiosity, need for admiration, and lack of empathy

- **Paranoid**—A pattern of distrust and suspiciousness such that others' motives are interpreted as malevolent

- **Schizoid**—A pattern of detachment from social relationships and a restricted range of emotional expression

- **Schizotypal**—A pattern of acute discomfort in close relationships, cognitive or perceptual distortions, and eccentricities of behavior

As you scan the list, you may recognize patterns you know by their colloquial names. Some of those names, such as *borderline* and *paranoid*, are the same as the clinical ones, while others are more colorful. For example, we use *sociopath* or *psychopath* for *antisocial; wallflower* for *avoidant; control freak, detail queen, workaholic, perfectionist,* or *bean counter* for *compulsive; clinger* for *dependent; drama queen* for *histrionic; egotist* or *narcissist* for *narcissistic; loner* for *schizoid;* and *weirdo* for *schizotypal*. But unlike the everyday words, which are used loosely and inconsistently, the DSM defines the Top Ten more carefully on the basis of clinical observations of enduring patterns of behavior. It also includes criteria for deciding how adaptive or maladaptive a pattern may be in a particular person. Those who are judged to be sufficiently impaired or distressed by an extreme and inflexible form of one or more of these patterns are said to be suffering from a personality disorder.[3]

In thinking about the Top Ten, it is important to recognize that, unlike the banana example, these are not clearly circumscribed natural categories.[4] Instead, they're more like the dimensional (graded) words used for traits than like the categorical (yes/no) words used for fruits: You can be more or less compulsive, but either you're a banana or you're not. Furthermore, detecting signs of one or more of these patterns need not be a cause for concern. The significance of mild or moderate versions must be judged on a case-by-case basis.

Nevertheless, what makes the Top Ten useful in everyday life is that they are a convenient way to focus attention on these common patterns. Although many of us have versions of these patterns that do us more good than harm,[5] their frequent association with difficulties in personal relationships or self-control makes it worth being on the lookout for them. Such awareness is particularly valuable when you're trying to figure out what's bothering you about someone and what to do about it.

In the rest of this chapter, I flesh out pictures of each of the Top Ten. But instead of asking you to memorize lists of characteristics, I follow the lead of psychologists such as Paul Costa, Thomas Widiger, and Robert McCrae, who use high or low rankings on facets of the Big Five to describe them.[6] This approach builds on what you have already learned about the structure of personality. It also gives you a way to make a combined assessment of a person's Big Five profile and potentially troublesome patterns with a single framework. To get started, let's consider two patterns on the very low end of Extraversion.

Very Low E: Two Eccentric Loners

All of us know people who like to be alone. But few of us have had much experience with those who are at the very bottom of the Extraversion scale because such outliers are so good at keeping to themselves. To give you an example of what a low E pattern feels like, I've excerpted a self-description that a student named Noitrix posted on Yahoo!:

> I have been thinking recently [whether] my life is so unnatural/weird if I compare myself to others, but for me, my life isn't weird or strange at all.

> I mean, I have never really had an interest in making friends. I had only 2 friends in my entire life, but I have no one at the moment. I do not feel loneliness or sadness or anything like that. For me, loneliness as a feeling does not exist because I always wanted to be a loner.

> In my free time, I don't go anywhere. I don't have any friends, and I don't want anyone, really. I don't want to be even with my family. At school, I don't talk to anyone. I have no desire to get close to anyone; in fact, I love to be alone. I don't know how it is possible, but I'm not attracted to girls—but I'm not attracted to boys, either. Never had a girlfriend because I never

wanted [one] because I find it pointless/useless. I
don't think I'll ever fall in love. I feel like I'm asexual.

I don't really care what people say about me. I don't
feel anything when someone praise[s] or criticize[s]
me. Also, I avoid eye contact when I meet strangers.

If I have to spend a lot of time with others, I feel like
they suck out life energy from me and I need to spend
a great deal of time alone in order to regenerate. I hate
rumors, I hate gossip, and I hate small talk.

My only goal in life is to achieve my dreams. Everything else is meaningless. Friendship/love doesn't
mean anything for me.

Noitrix's description of himself fits well with the schizoid pattern described in the DSM, which lists the following
characteristics: "neither desires nor enjoys close relationships,
including being part of a family; almost always chooses solitary activities; has little, if any, interest in having sexual experiences with another person; takes pleasure in few activities;
lacks close friends and confidants other than close relatives;
appears indifferent to the praise and criticism of others; shows
detachment or little emotion." But if you think about this pattern in the context of the Big Five, you will see that it can also
be described almost completely in terms of very low rankings
on all six facets of Extraversion: low warmth, low gregariousness, low assertiveness, low activity, low excitement-seeking,
and low positive emotionality. So all of Noitrix's unusual characteristics may simply be a reflection of his place at the bottom
of the spectrum of E.

Considering Noitrix's odd behavior in this way not only gives you a different way of understanding him, but it also helps you distinguish his schizoid pattern from an even odder low E pattern called schizotypal. Unlike schizoids, schizotypals are not just indifferent to people. They also actively dislike them, a sign of low A; feel anxious in their presence, a sign of high N; and have a highly idiosyncratic way of thinking about the world, which can be taken as a sign of high O. Because of this combination, schizotypals don't just keep a low profile. They can be flagrantly eccentric.

A notable example is Bobby Fischer, a misanthropic recluse who was forced into the public eye because he was one of the greatest chess players of all time. But despite his great talent, Fischer's contempt for almost everyone offended even his most ardent fans. To make things worse, his frequent expression of bizarre ideas, including a vicious hatred of Jews and Americans, alienated him further. Although he remained a legend in the chess world, Fischer dropped out of sight in his thirties and lived the rest of his life as a vagrant. When he briefly surfaced immediately after the September 11, 2001, attack on the World Trade Center, it was to announce on a Philippine radio station: "This is all wonderful news ... I applaud the act. ... F--k the U.S. I want to see the U.S. wiped out."[7]

Not all schizotypals are as blatantly odd as Bobby Fischer. Some are content to live in an unconventional way without antagonizing others. But their eccentric behavior is usually obvious enough to distinguish them from schizoid loners such as Noitrix who simply want to keep to themselves.

Very High E: The Disquietude of Histrionics

The potentially troublesome part of the Extraversion scale is not restricted to the low end. There's also a high E pattern called histrionic. Unlike the schizoids, who may stay under your radar screen, histrionics tend to capture your attention because they are so eager to engage you.

Histrionics don't just get top scores on gregariousness, activity, excitement-seeking, and positive emotionality. There's also a prominent sexual quality to their Extraversion. Just as schizoids express their low E by a lack of interest in sex, histrionics express their high E through flamboyant sexual expression. In the DSM, two outstanding characteristics of this pattern are: "interaction with others is often characterized by inappropriate sexually seductive or provocative behavior" and "consistently uses physical appearance to draw attention to self."

But unlike the schizoid pattern, which is largely limited to E, the histrionic pattern also includes notable rankings on the rest of the Big Five. Histrionics tend to be naively high on trust, a facet of A; high on impulsivity, a facet of N; high on romantic fantasy and feelings, facets of O; and low on self-discipline and deliberation, facets of C. In addition to their seductiveness, the DSM emphasizes their theatricality, suggestibility, and nonanalytical way of thinking.

It's not hard to find public figures who fit this picture and show business is a good place to start. Marilyn Monroe is a fascinating example because her strong desire to call attention to her body was already apparent when she was a little girl. Gloria Steinem's biography describes Marilyn's report of a recurrent childhood impulse to take her clothes off in church: "I wanted desperately to stand up naked for God and everyone else to see. I had to clench my teeth and sit on my hands to keep myself from undressing."[8] Marilyn's subsequent promotion of herself as a sex symbol, the ease with which she moved in and out of sexual relationships, and her exaggerated emotions off-stage all fit the histrionic pattern.

This pattern is also easy to spot in Hollywood men such as Marlon Brando with whom Marilyn had an affair. He, too, projected sexuality, but of a masculine type. While Marilyn was titillating the guys in *Gentlemen Prefer Blondes,* Brando was turning on the ladies in *A Streetcar Named Desire* with the pouting and emotional swings of a bad boy who could easily get out of control. As with Marilyn, this was not just acting. Brando, too, displayed this pattern off-camera, and his difficulties with studio bosses and directors were as recurrent and authentic as hers.

Of course, you don't have to be a movie star to be histrionic. Dramatic and physically demonstrative people are hardly rare. They are often irresponsible, irrational, and shockingly outgoing. But many attract a personal audience that finds them exciting and a great deal of fun.

Low A Patterns: Paranoids, Narcissists, and Antisocials

Three patterns of unusual A are in the Top Ten— paranoid, narcissistic, and antisocial—and all are at the low end. This doesn't mean that low rankings on A are necessarily troublesome. In fact, many people who rise to the top of their fields have prominent versions of one or more of these three patterns. Nevertheless, clinicians have focused their attention on them because extreme versions may be self-defeating, frequently invite retaliation, and bring grief to others.

Thinking of these patterns together is useful because all of them include low rankings on three of the facets of A. People with each of these patterns tend to be selfish rather than generous, combative rather than cooperative, and heartless rather than compassionate. What distinguishes the three patterns is a particularly low ranking on at least one other facet of A. Paranoids are suspicious rather that trusting, narcissists are arrogant rather than modest, and antisocials are deceptive rather than straightforward.

Of these patterns, the paranoid one is the easiest to spot because those who express it are often outspoken about their distrust and dislike of others. Being so convinced of other people's malevolence, they justify their contempt, combativeness, resistance to criticism, and tendency to bear grudges as legitimate defenses. They also tend to be cold and detached, signs of low E; dogmatic and insistent on their strongly held opinions, signs of low O; and easily angered, a sign of high N.

Although this pattern is not a prescription for popularity, it can be skillfully employed in vocations that require litigiousness and skepticism about human motives. Ralph Nader, for example, put it to good use in his brilliant career as a public advocate. Starting with a relentless campaign to uncover chicanery in the automobile industry, which forced the production of safer cars, he later turned his attention to other areas of corporate and government incompetence and corruption. For many years, he and his Nader's Raiders spearheaded important reforms.

But Nader's success as a crusader was not just fueled by paranoia. His ability to attract support for his populist movement was energized in part by the confidence and need for admiration that come with the narcissistic pattern.[9] Moderate versions of this pattern are common among inspiring leaders. But some go too far. The pattern becomes particularly troublesome if it expands into arrogant grandiosity that impairs judgment.

In Nader's case, the grandiosity was hard to miss in his 2000 campaign for president and its aftermath. Arguing that the two other candidates, Al Gore and George W. Bush, were as indistinguishable as "Tweedledee and Tweedledum ... so it doesn't matter which you get," Nader claimed that he was the only worthy candidate. When many of his early supporters urged him to drop out because he had no chance of winning and was pulling too many votes away from Gore, whom they preferred, Nader refused to get out of the limelight. And when Gore lost Florida by a few hundred votes—and, with it, the election—Nader wouldn't even consider the possibility that he had made a mistake. Instead, he was so pleased with himself that he wrote a book, *Crashing the Party*,[10] in which he exulted in his mischief and continued to insist that only he should have been elected.

Even before the 2000 campaign, the paranoid and narcissistic patterns that fueled many of Nader's successes had already gotten out of hand. We now know that the suspiciousness that helped him defeat outsiders also turned him against his colleagues at the first hint of disloyalty. And we know that the narcissistic traits that attracted dedicated crusaders to his early causes became justifications for exploitation and mean retaliation if they didn't follow him blindly. Lisa Chamberlain summed this up in *The Dark Side of Ralph Nader*:

> Dozens of people who have worked with or for Nader over the decades have had bitter ruptures with the man they once respected and admired. The level of acrimony is so widespread and acute that it's impossible to dismiss those involved as disgruntled former employees...his own record, according to many of those who have worked closely with him, is characterized by arrogance, underhanded attacks on friends and associates, secrecy, paranoia and mean-spiritedness—even at the expense of his own causes.[11]

The narcissistic pattern has other dark sides. One of the most common is the taking of unnecessary risks because of a sense of invulnerability, and many narcissists self-destruct because of such errors of judgment. Napoleon's invasion of Russia is an example.

But not all narcissists feel invulnerable. Many who lack the talent to be truly successful devote their energies to maintaining the illusion of superiority. To puff themselves up, they fantasize about a brilliant future, brag about their smallest achievements, and try to increase their status by putting others down. Nevertheless, such vulnerable narcissists[12] are easily crushed by even the smallest hints of criticism.

The great need of narcissists to feel high on the pecking order distinguishes them from people with a related low A pattern that the DSM calls antisocial and that other experts call psychopathic or sociopathic.[13] Like narcissists, antisocials are deceptive exploiters who lack empathy. But unlike narcissists, who are eager for admiration, most antisocials are not particularly interested in praise from others. Their cool indifference shows up in very low rankings on self-consciousness, vulnerability, and anxiousness, facets of N. In fact, their ability to experience negative emotions may be so low that they are incapable of feeling guilt or remorse and show no signs of conscience. Many of them also rank low on dutifulness and deliberation, facets of C, and are high on assertiveness and excitement-seeking, facets of E.

Considering how much damage antisocials can do, you might think that we would constantly be on the lookout for them. Yet they are remarkably easy to miss. One reason we may be so blind to them is that most of us find it hard to believe that such people really exist. Furthermore, they tend to be such glib deceivers that we may keep dismissing the evidence that they're conning us, even if we keep catching them in the act. Robert Hare, an expert on the psychopathic pattern, remembers how he, too, used to be fooled by them. When he talks about such people at a party, he often gets responses like, "You know, I never realized it before, but the person you're describing is my brother-in-law."[14]

Bernard Madoff,[15] who operated a massive Ponzi scheme for 20 years, is a good example. When the scheme was finally exposed, many of those he had been swindling for two decades just couldn't believe it. "How could such a nice man do such a terrible thing?" "How could he keep screwing his closest friends, the people who kept trusting him?" Yet here he was, a seeming pillar of the community who had gone on lying and stealing for years in the face of repeated investigations—shameless, remorseless, unconstrained by conscience.

Madoff didn't fool only gullible clients. He also fearlessly faced down officials of regulatory agencies who were trained to detect fraud. Even when an economic meltdown led to massive withdrawals that finally exposed his scheme, Madoff remained confident that he could cut a deal—so confident that he didn't bother to consult his lawyer before he confessed.

O.J. Simpson, another famous antisocial, shared Madoff's belief that he could get away with anything. After he was accused of murdering his wife, Nicole, and her friend, Ron Goldman, and with a trail of damaging evidence against him, Simpson stayed cool. His brazen demeanor during his criminal trial, and the ease with which he played with the murder glove, helped persuade the jury that he wasn't guilty.

Even Simpson's subsequent conviction in the civil trial didn't faze him. And instead of putting the whole thing behind him after that verdict was announced, Simpson decided to write a book, called *If I Did It*, an in-your-face virtual confession that further illustrates the callousness of antisocials. By describing the details of the way he *might* have committed the murders, Simpson could take pleasure in taunting the families of his victims while still claiming innocence.[16]

Such sadistic pleasure is illustrated even more vividly in Javier Bardem's Academy Award–winning portrayal of a psychopathic killer in *No Country for Old Men*. In a particularly chilling scene early in the film, we see him toying with the hapless attendant at a gas station, who is quickly transformed from friendly to terrified. The man becomes increasingly bewildered by Bardem's subtle threats and his unwillingness to back off in the face of signs of conciliation. Only because of the luck of a coin flip does the attendant escape with his life.

Bardem's fictional character is, of course, an extreme version of this pattern, a killing machine who loves his work. Simpson's version is more moderate, and many components of the pattern served him well in his outstanding football career. Were it not for the close scrutiny that followed Nicole's murder, his great athletic achievements might have allowed him to continue to get away with a lot of antisocial behavior. From Robert Hare's perspective, Bardem's character is a good example of a full-blown psychopath, whereas Simpson might have been considered a "subcriminal psychopath."[17]

Many other antisocials who reach high positions are even more skillful at covering their tracks. And such people are not rare. Surveys show that about 4% of Americans, mostly male,[18] fit the antisocial picture described in the DSM. So if someone you know shows signs of it, don't dismiss it out of hand. It's worth staying on the lookout for additional evidence.

Very High C: Compulsives

Although the antisocial pattern is fairly common, it is not the most widespread of the Top Ten. The compulsive pattern holds that record. A recent survey found troublesome forms of this pattern in about 8% of American adults,[19] both women and men.

A major distinguishing feature of the compulsive pattern is high scores on all facets of Conscientiousness. But is there anything wrong with this? Aren't competence, orderliness, dutifulness, self-discipline, deliberation, and achievement-striving exactly what our parents kept encouraging? Aren't they crucial ingredients of success? So what's the difference between the adaptive pattern of high C that is associated with healthy achievement and the potentially troublesome pattern of high C called compulsive?

As with all troublesome patterns, it is a matter of degree. For example, Theodore Millon describes gradations of the component he calls perfectionism that range from adaptive ("I take pride in what I do"), to disordered ("I can't stop working on something until it's perfect, even if it already satisfies what I need it for"), to severely disordered ("because nothing is ever good enough, I never finish anything").[20] DSM's description includes other signs of maladaptive perfectionism: "is preoccupied with details, rules, lists, order, organization or schedules to the extent that the major point of the activity is lost"; "is overconscientious, scrupulous, and inflexible about matters of morality, ethics, or values"; and "is reluctant to delegate tasks or to work with others unless they submit to exactly his or her way of doing things."

But it isn't just the degree of high C that accounts for the troublesome forms of this pattern. After all, many super-achievers express top scores on C in an adaptive way. A distinguishing feature of maladaptive high C is that it tends to be associated with high Neuroticism, especially high anxiousness and vulnerability. Unlike healthy high Cs, whose hard work may be rewarded by the joy of achievement, maladaptive ones take little pleasure in what they do. Instead, they are motivated by the intense desire to avoid mistakes, and their distress can become unbearable if they don't do things in a certain way. Nobody knows why they choose a slavish commitment to hard work as their main tactic for warding off negative emotions. But whatever the reason, they are prisoners of perfectionism, locked in a pattern that brings no happiness to them or to anyone else.

High N Patterns: Avoidants, Dependents, and Borderlines

High Neuroticism, which brings so much grief to maladaptive compulsives, is also responsible for the distress that accompanies the three remaining Top Ten patterns: avoidant, dependent, and borderline. But unlike compulsives, whose N flares up if they deviate from their rigid ways of doing things, the high N of these three patterns is mainly expressed as a feeling of vulnerability in social situations and relationships. Because of this common feature, people who have prominent versions of one of them may also have signs of the others.

The easiest to spot are the avoidants because they are uncomfortable in groups. But unlike schizoids, with whom they are sometimes confused, many avoidants are actually interested in socializing. The reason they hang back is their worry that they are personally unappealing, which makes them afraid of being embarrassed and rejected.

This difference between avoidants and schizoids shows up in their scores on Neuroticism. Avoidants are particularly high on self-consciousness and vulnerability, which together drive their dread of disapproval, whereas schizoids couldn't care less if other people look down on them. The DSM also emphasizes broader aspects of the avoidant pattern, such as "is unusually reluctant to take personal risks or to engage in any new activities because they may prove embarrassing."

But many avoidants do find a way to become engaged with people, and some may even rise to positions of prominence, despite their fears and inhibitions. A good example is William Shawn, who edited the *New Yorker* for 35 years. His son Allen described his father's avoidant pattern, and his way of coping with it, in *Wish I Could Be There:*

> He had what might in retrospect seem like a strong streak of social phobia. In addition to avoiding crowded places and always sitting on the aisle or near an exit in any theater or concert hall, he avoided most parties and get-togethers. I don't remember his instigating a party of his own. Rather he seemed a somewhat reluctant, passive participant in a social gathering, though he usually ended up being the quiet epicenter of the event. He would walk into even his own living room rather tentatively if it contained guests, looking cheerful and ruddy-faced but also hanging back. Though he spent all day with people, they seemed to astonish him. His respect for the complexity and mystery of others was part of what made him a deep person, but it also expressed some inner fear...
>
> He was famously shy, preferring to speak to individuals rather than to a group... He had, I believe, no

actual fear of anyone and in a sense was profoundly sociable. He just needed certain conditions in which to reveal his sociability, just as he needed certain conditions in which to assert himself, to be spontaneous, and to reveal his pride in himself...[21]

This ability of some avoidants to assert themselves is not shared by people with another high N pattern, called dependent. Instead of fighting against their deep sense of insecurity, they seek out stronger people as potential protectors. The DSM's description of this pattern includes: "has difficulty making everyday decisions without an excessive amount of advice and reassurance from others"; "needs others to assume responsibility for most major areas of his or her life"; "feels uncomfortable or helpless when alone because of exaggerated fears of being unable to care for himself or herself"; and "is unrealistically preoccupied with fears of being left to take care of himself or herself."

This group of vulnerable people can take this path because they are also relatively high in Agreeableness. Believing that there are many generous people who won't take advantage of them, they are not ashamed to admit their limitations. Instead, they feel free to express their eagerness to ingratiate themselves, in the hope that their trust will be reciprocated and that they will find a loving companion they can rely on.

It sometimes works. If dependents get themselves into a stable relationship, their N may stay under the surface while only their Agreeableness shines through. But a troublesome outcome is not unusual because dependents often overestimate the commitment of their partner. When the honeymoon is over, they may become clinging and demanding, and fear of abandonment may overwhelm them.

Such fear of abandonment is also a prominent feature of the borderline pattern, an extreme expression of Neuroticism. Borderlines have high scores on all facets of N: anxiousness, angry hostility, depressiveness, self-consciousness, impulsivity, and vulnerability. To make things worse, they also have low scores on trust and compliance, facets of A, and a low score on deliberation, a facet of C. But it is the N that stands out, and its expression may include both angry disappointment and cling-ing dependency, the combination of *I Hate You, Don't Leave Me*,[22] which is the title of a popular book about this pattern.

The description of troublesome forms of this pattern in the DSM begins with three signs of such turbulence: "fran-tic efforts to avoid real or imagined abandonment"; "unstable and intense interpersonal relationships alternating between extremes of idealization and devaluation"; and "unstable self-image or sense of self." The picture, then, is one of intense interpersonal needs, strong attachment, and fears of betrayal. Prone to loneliness, people with this pattern often seek com-fort from sexual promiscuity and illegal drugs.

Despite its extreme nature, don't be surprised if this descrip-tion reminds you of someone you know. Researchers from the National Institutes of Health detected troublesome versions of the borderline pattern in about 5% of the Americans that they examined in face-to-face interviews. And despite the widely held belief that most borderlines are women, the researchers found that this tumultuous pattern is also common among men.[23]

Milder versions of the borderline pattern also exist that Millon considers "on a continuum with normality"[24] and that Oldham and Morris call "the mercurial style."[25] Such people are eager to be involved in romantic relationships, seek intense closeness, and are easily hurt if these feelings are not enthusiastically reciprocated at all times. But their breaking up and making up is more modulated, their moods are less volatile, and their view of their relationships is more realistic.

Opinions of Self and Others

Now that I've described these patterns in terms of the Big Five, let's turn to another way to conceptualize them that I also find helpful. This method is based on research by psychiatrist Aaron T. Beck, who studied the thought processes of people with troublesome personalities as a guide to their treatment. His approach, called cognitive therapy, is designed to help his clients identify and re-examine the ways of thinking that get them into trouble. In developing this form of psychotherapy, Beck and his coworkers identified two highly informative thought processes: a person's opinion of himself and his general opinion of others. They also found that particular opinions of this kind are characteristic of each of the Top Ten.[26]

In looking for signs of such potentially troublesome thought processes, I begin by restricting my attention to the opinions of themselves. And instead of trying to sort through ten alternatives, I've lumped them together into four categories. Two of them project positive self-images: "I'm special" and "I'm right." The other two are more negative: "I'm vulnerable" and "I'm detached."

These four kinds of opinions of self are probably familiar because they frequently come up when we gossip about people. For example, we may say, "She's so full of herself" (special), "He's so self-righteous" (right), "She's so insecure" (vulnerable), or "He's a real loner" (detached). If these or similar statements seem to fit the person you have in mind, you can refine your assessment by seeing how well it matches up with the characteristics summarized in Table 2.1.[27]

TABLE 2.1   Top Ten Patterns: Opinions of Self and Others

	Pattern	Self	Others
I'm Special	Narcissistic	Entitled	Inferior
	Histrionic	Glamorous	Seducible
	Antisocial	Unconstrained	Suckers
I'm Right	Compulsive	Competent	Slackers
	Paranoid	Righteous	Malicious
I'm Vulnerable	Avoidant	Unappealing	Demeaning
	Dependent	Needy	Supportive
	Borderline	Unstable	Inconsistent
I'm Detached	Schizoid	Self-sufficient	Unrewarding
	Schizotypal	Magical	Untrustworthy

The three ways of thinking "I'm special" have some similarities and clear differences. Narcissists believe they are superior and above the rules. They expect others to admire them and to offer them the special treatment they are convinced they deserve. Histrionics also expect admiration, but mainly for their glamour. And, unlike narcissists, histrionics don't see others as inferior. Instead, they view them as potential targets for seduction. Antisocials share the sense of superiority of narcissists, but they are mainly interested in taking advantage of people rather than being admired by them. They believe that what makes them special is that they are unconstrained by social conventions. This allows them to deceive and exploit the suckers of the world. It also allows many of them to keep getting away with it because they are so good at hiding their true aims.

The two ways of thinking "I'm right" are also fairly easy to tell apart. Compulsives consider themselves competent and committed to excellence. They consider others to be self-indulgent slackers who should work harder and follow the rules. Paranoids may be even more self-righteous. But they also feel misunderstood, despite what they consider to be their noble intentions. Instead of dismissing others as irresponsible, they are wary of them as malicious antagonists.

The three ways of thinking "I'm vulnerable" each include a particular version of the belief "I'm not good enough" and can also be distinguished by their very different views of others. Avoidants are particularly concerned that others see through them, recognize their ineptness, and are eager to put them down. To prevent embarrassment, they keep a low profile. Dependents also feel inept but are not ashamed to reach out to people who may take care of them. Borderlines, the most flagrantly troublesome, have an unstable view both of themselves and of others. They are acutely aware of their limitations but also cling to the belief that they are adored. They swing between a positive view of people they become attached to, whom they consider loving and perfect, and the negative view that they are in constant danger of being betrayed and abandoned by them.

The two ways of thinking "I'm detached" also include very different views of self and others. Schizoids have a sense of self-sufficiency that reflects their ability to take care of themselves, and they stay away from others because they find relationships messy and unrewarding. In contrast, schizotypals have a sense of self-sufficiency because they live a fantasy world that they prefer to the real one, and their main reason for staying away from others is that they suspect them of being untrustworthy.

Traits, Patterns, and People

Considering the Top Ten as both a pattern of traits and a pattern of thoughts underscores their value as a vocabulary for discussing people and making predictions about them. So if you identify someone's boss as narcissistic, you can better understand why he demoralizes an avoidant employee but angers a paranoid one. And if you identify a friend as histrionic, you can better understand why she is a sitting duck for a smooth-talking antisocial.

Useful though this may be, it is important to remember that the Top Ten are not sharply defined natural categories. For example, there are all kinds of narcissistic bosses. Nevertheless, identifying someone as narcissistic, using the characteristics I've described, still communicates real content that further observation and analysis can either confirm or reject. The same is true for the other patterns on the list.

When viewed in this way, the hunch that a person has a potentially troublesome pattern can help you organize your observations about his notable Big Five tendencies. In the case of narcissism, it might first focus your attention on the facets of low A that tipped you off. If your hunch is confirmed, Conscientiousness might be the next one to consider: High C can propel people with the narcissistic pattern to great achievements, while low C may move them in an antisocial direction. Rankings on N, O, and E also change the complexion of this pattern in many different ways. So building a Big Five assessment around an initial hunch about someone can be more fruitful than just going through the list of traits without a working hypothesis.

As you learn to think of people in terms of both their tendencies and patterns, you will not only start seeing them more clearly, you will also become increasingly aware of the great variety of human personalities. This raises questions about the origins of these many variations, questions that I turn to in the following chapter.

Practical Summary

In this chapter I've shown you how to take the next step in your assessment of a personality by looking for troublesome patterns.

The ten patterns I described come from psychiatry's diagnostic manual, which was designed to identify them in their full-blown and maladaptive forms, called personality disorders. But these patterns also exist in milder forms that need not be maladaptive. In assessing personalities you should be on the lookout for both the milder and full-blown versions. Your findings will then take their place in the overall personality picture you are developing.

REVIEWING THE TOP TEN

To review these patterns, here, again, are brief summaries. To make them more vivid, I've included descriptive phrases from the DSM. In rethinking each of them, please ask yourself if it calls to mind the behavior of someone you know:

- **Antisocial**—A pattern of disregard for and violation of the rights of others. "Deceitfulness as indicated by repeated lying, use of aliases, or conning others for personal profit or pleasure. Lack of remorse as indicated by being indifferent to or rationalizing having hurt, mistreated, or having stolen from another."

- **Avoidant**—A pattern of social inhibition, feelings of inadequacy and hypersensitivity to negative evaluation. "Shows restraint within intimate relationships because of fear of being shamed or ridiculed."

- **Borderline**—A pattern of instability in interpersonal relationships, self-image, and emotions and marked impulsivity. "Unstable and intense interpersonal relationships alternating between extremes of idealization and devaluation."

- **Compulsive**—A pattern of preoccupation with orderliness, perfectionism, and control. "Is overconscientious, scrupulous, and inflexible about matters of morality, ethics and values. Is reluctant to delegate tasks or to work with others unless they submit to exactly his or her way of doing things."

- **Dependent**—A pattern of submissive and clinging behavior related to an excessive need to be taken care of. "Needs others to assume responsibility for most major areas of his or her life."

- **Histrionic**—A pattern of excessive emotionality and attention seeking. "Interaction with others is often characterized by inappropriate sexually seductive or provocative behavior. Is uncomfortable in situations in which he or she is not the center of attention."

- **Narcissistic**—A pattern of grandiosity, need for admiration, and lack of empathy. "Is interpersonally exploitative, i.e., takes advantage of others to achieve his or her own ends. Is preoccupied with fantasies of unlimited success, power, brilliance, beauty, or ideal love."

- **Paranoid**—A pattern of distrust and suspiciousness such that others' motives are interpreted as malevolent. "Reads hidden, demeaning, or threatening meanings into benign remarks or events."

- **Schizoid**—A pattern of detachment from social relationships and a restricted range of emotional expression. "Neither desires nor enjoys close relationships, including being part of a family."

- **Schizotypal**—A pattern of acute discomfort in close relationships, cognitive and perceptual distortions, and eccentricities of behavior. "Behavior or appearance that is odd, eccentric, or peculiar."

APPLYING THE TOP TEN

Having gone through this list, please consider the following:

1. How many of these patterns did you identify among people you know?

2. In each case, is the person troubled by it? Are you?

Judging whether a pattern is troublesome depends, of course, on your point of view. For example, some of us avoid close relationships with perfectionists and are only comfortable with people who are laid back. But many of us don't seem to mind this pattern, and others are strongly attracted to compulsive partners.

THREE WAYS TO NOTICE TROUBLESOME PATTERNS

1. **By pattern recognition**—The easiest way to notice troublesome patterns is to be on the lookout for their characteristic features. Becoming aware of them is worthwhile not only in the course of systematic personality assessments, but also in your everyday life. If you pay attention, you'll find that many patterns are obvious. But even experts may miss some. This is especially true of the antisocial pattern, which is often successfully hidden by those who express it.

2. **By focusing on attitudes about self and others—**
Another way to detect troublesome patterns is to keep your eyes open for characteristic attitudes about self and others:

> **Antisocial.** I'm Unconstrained/Others are Suckers
>
> **Avoidant.** I'm Unappealing/Others are Demeaning
>
> **Borderline.** I'm Unstable/Others are Inconsistent
>
> **Compulsive.** I'm Competent/Others are Slackers
>
> **Dependent.** I'm Needy/Others are Supportive
>
> **Histrionic.** I'm Glamorous/Others are Seducible
>
> **Narcissistic.** I'm Entitled/Others are Inferior
>
> **Paranoid.** I'm Righteous/Others are Malicious
>
> **Schizoid.** I'm Self-sufficient/Others are Unrewarding
>
> **Schizotypal.** I'm Magical/Others are Untrustworthy

3. **As part of a Big Five assessment—**Troublesome patterns may also show up while you're doing a Big

Five assessment. For example, a schizoid pattern is suggested by low rankings on all facets of E.

P AND TROUBLESOME PATTERNS

Having reviewed these patterns from several perspectives, it's time to apply them to P by considering the following questions:

1. Does P show characteristics of any full-blown Top Ten patterns?

2. Does she have characteristics of any mild versions?

   If she does:

3. Did you notice it by pattern recognition? Because of her attitude about herself and others? By assessing her Big Five? In all these ways?

4. Do you believe any patterns you've observed, whether full-blown or mild, are troublesome enough to interfere with her relationships?

5. If you consider any of P's patterns troublesome, might they also play adaptive roles in her overall functioning?

6. Having thought through these questions, do you see the value of adding the Big Ten to a systematic personality assessment?

PART II

Explaining Personality Differences

Every night and every morn
Some to misery are born,
Every morn and every night
Some are born to sweet delight.

—William Blake, *Songs of Innocence*

THREE

How Genes Make Us Different

In considering the cast of characters I described in the previous chapter, you may have wondered how they got to be so different. If you're like most people, you probably assumed that their personality patterns were mainly caused by social circumstances and upbringing. But it's likely that you also toyed with another explanation that is becoming increasingly popular: genes.

The growing interest in the genetics of personality is reflected in its extensive media coverage. Consider, for example, this excerpt from a *New York Times* column about the genetics of excitement-seeking:

> Jason Dallas used to think of his daredevil streak—
> a love of backcountry skiing, mountain bikes and
> fast vehicles—as "a personality thing." Then he
> heard that scientists at the Fred Hutchison Cancer
> Research Center had linked risk-taking in mice to a
> gene. Those without it pranced unprotected along a
> steel beam instead of huddling in safety like the other
> mice. Now Mr. Dallas, a chef in Seattle, is convinced
> he has a genetic predisposition for risk-taking, a con-
> clusion that researchers say is not unwarranted, since
> the similar variations in human genes can explain

why people perceive danger differently. "It's in your blood," Mr. Dallas said. "You hear people say that kind of thing, but now you know it really is."[1]

What I find remarkable about this report is that Jason Dallas so readily accepts the idea that a gene that affects the personality of a mouse may also affect his own. Although there is no evidence that the gene in this study, neuroD2,[2] has anything to do with his love of excitement, Dallas has been primed to make this connection by the widely publicized findings that there is, in fact, a close relationship between mouse genes and human genes, and between mouse brains and human brains. Needless to say, there are also important differences. But as I show in this chapter, Dallas does have good reason to believe that his daredevil streak has some genetic basis, even though his neuroD2 may have nothing to do with it.

The belief that some personality traits are innate is hardly new. What is new is our growing understanding of the degree and nature of this genetic influence. In this chapter, I take you beyond the vague idea that genes affect personality, to a deeper conception of the role they play in making us who we are.

A New Foundation for Psychology

Charles Darwin, who revolutionized our understanding of the origins of personality differences, didn't begin with a particular interest in this subject. He was after something much bigger: the origin of *all* the differences among *all* living things. Of the clues that led him to that answer, the most revealing came from domestic animal breeding.

Dogs were especially informative. People already knew in Darwin's time that breeds as different as greyhounds and spaniels descended from the same wild ancestors. Darwin also understood that their selective breeding depended on the transmission of inherited characteristics from parents to pups and that new breeds arose "by the careful selection of the individuals which present the desired character." Furthermore, the creation of strikingly different breeds was a gradual affair, accomplished through a succession of little steps. As Darwin explained this in 1859 in *Origin of Species*:

> [W]hen we compare the many breeds of dogs, each good for man in different ways ... we cannot suppose that all the breeds were suddenly produced as perfect and useful as we now see them; indeed, in many cases, we know this has not been their history. The key is man's power of accumulative selection: nature gives successive variations; man adds them up in certain directions useful for him. In this sense he may be said to have made for himself useful breeds.[3]

Once Darwin recognized that the creation of dog breeds depends on the breeder's selection of heritable variations, it occurred to him that nature does the same thing: It selects those heritable variations—spontaneous modifications of genes, now called mutations—that are advantageous in the wild. This process of natural selection ensures that desirable mutations are passed on from generation to generation and may eventually become stable features of the species.

A good example is a mutation in a gene, SLC24A5, which controls the deposit of melanin, a black pigment. What makes this mutation so interesting is that it caused a dramatic change in the color of human skin, from black to white. In sunny Africa black skin is favored to block harmful ultraviolet rays while still allowing enough through to stimulate the skin's production of vitamin D. This explains why the native African population has SLC24A5 genes that provide lots of black pigment. But in regions far from the equator, where sunlight is scarce, a mutation that inactivates this gene[4] took over because the pale skin that results lets through more of the limited light to make vitamin D.[5] As with the evolution of many other human differences, this one became prevalent through accidental DNA mutations and natural selection based on adaptation to specific environmental conditions.

Darwin wasn't in a position to provide such a persuasive illustration. But this didn't stop him from extending his idea from biology to psychology. It was clear to him that selective breeding affected not only physical characteristics, but also behavioral ones. For example, breeders have selected dogs not only for their shape and size, but also for their skills, such as herding or pointing, and for personality traits such as agreeableness or aggressiveness. So why wouldn't natural selection of behavioral traits also increase fitness in the wild? By the end of *Origin of Species,* Darwin was sufficiently convinced of this to predict that "In the distant future...psychology will be based on a new foundation, that of the necessary acquirement of each mental power and capacity by gradation." To put this in modern terms, Darwin predicted that our understanding of psychology would one day rely on knowledge of the genetic variations that affect behavior.

But Darwin was initially reluctant to extend this prediction from animals to people. The mere hint that physical features of humans had animal origins would cause him trouble enough. He would, for some time, leave human psychology to others.

Experiments of Nature

The man who first took up the challenge was Darwin's cousin, Francis Galton, whom you met in Chapter 1, and he was willing to take it even further. It seemed to Galton that if the characteristic behaviors of a species are inherited, the behavioral differences between individual people—our distinctive intellectual abilities and personality traits—might also be inherited.

Such variations in human talents and traits were already of great personal interest to Galton. A precocious child who was proud of his intelligence and achievements, he had long believed that both he and Charles Darwin had inherited their special gifts from their common grandfather, the distinguished physician and scientist Erasmus Darwin. But Galton was aware that his family also provided him with a privileged upbringing that fostered whatever gifts he inherited by placing him "in a more favourable position for advancement than if he had been the son of an ordinary person."[6] So was he gifted because of favorable heredity or favorable upbringing?

To address this question, Galton turned to an experiment of nature: twins. Galton knew that some twins looked so much alike that they were probably genetically identical, whereas others were no more similar than siblings born at different times. Because both identical and fraternal twins were usually raised together by their parents, the members of each pair would have a comparable upbringing. If he found that the behavior of identical twins was more similar than that of same-sex fraternal twins, this would support his hunch that greater genetic similarity leads to greater behavioral similarity.

In 1875, Galton reported that 35 sets of identical twins showed much greater behavioral similarities than 20 sets of fraternal twins, which he took as support for the importance of heredity. In "The History of Twins as a Criterion of the Relative Powers of Nature and Nurture," he announced, "[T]here is no escape from the conclusion that nature prevails enormously over nurture."[7] His observations that adopted children of gifted adoptive parents are no more gifted than ordinary children, even though they are provided with a privileged environment, also supported this conclusion.[8] This was the first use of another natural experimental approach—adoption—in assessing the role of inheritance and upbringing.

Although Galton's ways of studying behavior were crude, his results were sufficiently persuasive to convince his most eminent critic. As Darwin wrote to him after studying some of Galton's publications, "I do not think I ever in all my life read anything more interesting and original—and how well and clearly you put every point! ... You have made a convert of an opponent in one sense, for I have always maintained that, excepting fools, men did not differ much in intellect, only in zeal and hard work."[9] While Darwin's praise was not wholly merited in its time, it was subsequently justified by more persuasive research using Galton's approach.

How Much of Our Personality Differences Is Heritable?

The biggest impediment to Galton's research is that he didn't know how to measure personality differences. He had tried to make objective assessments in his work with twins, but he was painfully aware that his methods weren't very good. Frustrated by these difficulties, Galton turned his attention to the inheritance of height, which he could measure accurately. His studies of the relationship between the heights of parents and their children led him to develop the formula for calculating correlations that I mentioned earlier, and that was later adapted to create the Big Five personality tests.

The Big Five tests are just what Galton had hoped for, and they are now routinely used to investigate the influence of genes on the personalities of identical and fraternal twins. In a typical study, each twin is given a Big Five test, and the scores are compared with those of the other twin. If genes influence these personality traits, both twins should have scores that are somewhat similar. But the similarities of pairs of identical twins, who share 100% of their genes (because they are derived from a single fertilized egg that split after conception), should be twice as great as the similarities of same-sex fraternal twins (derived from different eggs), who share only 50% of their genes.

This is just what researchers have found. For example, in a study using hundreds of subjects, the Extraversion scores of the two members of a pair of fraternal twins had an average correlation of 0.23 (on a scale of 0 to 1). In contrast, the two members of a pair of identical twins had an average correlation about twice as large, 0.48. The difference in correlations (0.48 − 0.23 = 0.25) is assumed to reflect the difference between having all the same genes (identical twins) and having half the same genes (fraternal twins). Therefore, this difference measures only half the effect of having all the same genes. To get the full effect, which geneticists call heritability, 0.25 is doubled to get 0.5, or 50%.[10] Studies of Agreeableness, Conscientiousness, Neuroticism, and Openness also found heritability to be around 50%.[11]

When the evidence for such substantial heritability of personality traits was first published, critics pointed out another possible explanation for the greater psychological resemblance of the identical twin pairs. Instead of resulting from genes alone, it might also result from the identical twins being treated more alike than the fraternal twins. Fortunately, the contribution of shared family environment can be evaluated through another experiment of nature: studying identical twins separated after birth and raised in different families.

Thomas Bouchard and his colleagues at the University of Minnesota did just that.[12] They tracked down more than 100 pairs of identical twins who had been raised apart and persuaded them to volunteer for a week of psychological testing. Many twins had been raised in very different environments, some in different countries and cultures, and their reunions attracted a great deal of media attention.

A certain pair of British identical twin girls would have been especially interesting to Galton because they addressed the issue of social privilege he had wondered about. One twin had been raised by an upper-class family, had attended exclusive schools, and spoke with a refined accent to prove it; the other had been raised by a lower-class family, had quit school at 16, and spoke like Liza Doolittle did before she met Henry Higgins. Yet their test scores were very similar. The same was true of the other sets of twins. As Bouchard summed it up, "[O]n multiple measures of personality and temperament, occupational and leisure-time interests, and social attitudes, monozygotic twins reared apart are about as similar as monozygotic twins reared together."[13] These and other family and adoption studies support the conclusion that personality traits are highly heritable.[14]

The studies with identical twins also tell us something else that should not be ignored: They challenge the assumption that the shared family environment of those raised together is responsible for some of their similarities. Were this the case, the scores of identical twins raised together should be more similar than those raised apart. But as Bouchard pointed out, they're not.[15] Scores of genetically unrelated children who were adopted and raised in the same family also show no effect of this shared environment.[16]

The lack of effect of a shared family environment on these measures of personality doesn't mean that parents are just part of the furniture. Studies indicate that parents do have some influence, but it is transmitted by their unique relationship with each child,[17] including each identical twin. The studies also indicate that most environmental influences on personality cannot be specifically attributed to interactions within the family.[18]

How Many Gene Variants Shape a Personality Trait?

So now that we know that genes do, indeed, have a big effect on personality differences, how do they do it? To answer this question, it's necessary to review a few facts about human genetics.

The total number of human genes is surprisingly small, about 20,000. Each is made from the four chemical building blocks of DNA—adenine (A), cytosine (C), guanine (G), and thymine (T)—strung together in a long chain whose precise sequence (such as AGACTCAAG ...)contains the instructions for manufacturing a particular protein. Each protein interacts with many others to build and maintain us. The major reason so few genes are sufficient for this complex task is that various combinations work together to control our biological and psychological functions. Furthermore, the actions of each gene and each protein can influence the actions of many others.

The main way genes interact is by turning the activities of each other on or off. To make this possible, each gene has a specialized piece of DNA, called a promoter, which serves as a dial to control the amount of the protein that gene makes. The dial can be turned up and down by internal or environmental signals that may work through controls in other regions of DNA. This process, called regulation of gene expression,[19] adjusts the amounts of the proteins that shape our bodies and minds.

Regulating the expression of a variety of genes in different cells helps explain how just 20,000 elements can give rise to such complexity. But it doesn't explain our heritable differences. These differences are explained by *variants* of the genes—modifications of the sequence of bases in their DNA or the DNA of their regulatory regions—that have accumulated in the collective human genetic repertoire, called the human genome. These structural modifications of DNA, which arose through random mutations, may cause big changes in the manufacture of a specific protein or the way it functions in the body. Some of the variants, such as those that influence skin color, are carried by billions of people. Others are rare. The combined effects of the assortment of gene variants that were handed down to each of us—our own personal selection from the human genome—define our genetic uniqueness.

But not all gene variants have such major and obvious effects as the small number that control human skin color. For example, hundreds of different genes[20] influence human height, which is about 80% heritable in well-nourished human populations, and each of these genes has a tiny effect. The same is true for highly heritable personality traits.

Persuasive evidence that personality traits reflect the joint action of multiple gene variants comes from selective breeding of mice. A notable example is John DeFries's classic study[21] of a seemingly simple mouse trait: the inclination to explore an unfamiliar and potentially dangerous territory. This personality trait is related to both the excitement-seeking and anxiety facets of the Big Five, in that high excitement-seeking would increase exploration while high anxiety would inhibit it. Together these facets would also influence the risk taking that is such a cherished part of Jason Dallas's personality.

To prepare for the experiment, DeFries randomly chose ten litters of mice and observed their behavior in a brightly lit large box called an open field. Mice prefer dim light and narrow spaces, but there are individual differences. Some mice froze in the open field, like a deer in the headlights, while others sniffed around and explored. Electronic sensors measured each mouse's behavior, recording the total distance it traveled in a six-minute period. After DeFries had scored each animal from the initial ten litters, he selectively bred the mice to raise two extreme lines. He began by mating the most active male and female from each litter. They became the founders of what I call the fearless (F) line. He also mated the least active pairs, who became the founders of what I call the anxious (A) line. DeFries then took another ten litters and mated a randomly selected male and female from each to be the founders of the control line. He repeated this process in each generation. Because mouse pregnancies take only three weeks, and pups become sexually mature in about three months, he was able to breed and evaluate 30 generations in the course of ten years.

The results were dramatic. After 30 generations, the average member of the F line roamed freely across the open field. In contrast, the average member of the A line huddled in a corner of the box. Members of the control line maintained their original modest level of exploration, which hadn't changed through 30 generations.

The other notable finding was that separation of the two lines was gradual, with steady increments from one generation to the next. When the open field behavior of the F line was plotted as a graph, it looked like that of a long-term growth stock that kept rising year after year over the ten-year period. In contrast, the pattern of the A line looked like the stock of a weak company in a failing industry, heading progressively downward until it hovered near zero. This pattern of gradual change has two implications: Variants of many genes together affect this personality trait; and the behavioral effects keep adding up as more of the relevant gene variants are selected in each generation. Direct analysis of the DNA of these mouse lines[22] confirms these conclusions.[23]

Genetic Thinking vs. Genetic Testing

As behavioral scientists were accumulating evidence that many gene variants work together to influence personality differences, geneticists were busily deciphering the complete DNA sequences of the human and mouse genomes and the structures of common gene variants. This provided the foundation for searching the entire genome for variants that influence personality traits in people and in mice. But progress with this genome-wide approach has been slow.[24]

Frustrated by this limited success, some researchers have taken a more focused approach. It was based on the knowledge that drugs such as Prozac and Ritalin affect personality and

that they do this by influencing the way serotonin or dopamine act in the brain. This raised the possibility that inherited
variations in genes that control certain actions of serotonin or
dopamine in the brain might be responsible for heritable personality differences. To look into this, researchers examined
variants of several dozen of these genes to see if they were correlated with scores on personality tests.

All the genes that they examined influence the emotional
circuits of the brain. Of these, the most widely studied, called
SERT, was singled out because it makes the serotonin transporter protein, Prozac's target. The SERT protein vacuums up
(transports) serotonin from the fluid around nerve cells that
are activated by frightening experiences so that the serotonin
can be used again. By controlling the amount of serotonin that
bathes these nerve cells, the SERT protein affects the intensity
of the emotional response. Therefore, it is easy to imagine how
variants of the SERT gene might influence the tuning of brain
circuits that control traits such as fearfulness.

To see whether the SERT gene affects personality, researchers focused on two common variants, one with a long promoter
and the other with a short one. Several studies have found
that groups of people who have two copies of the gene with
the long promoter, which makes more SERT protein, have a
slightly lower average score on Neuroticism.[25] Furthermore,
brain imaging studies indicate that if such people are shown
frightening pictures, they have less activation of the amygdala,
a brain structure involved in fear processing.[26] Taken together,
the studies suggest that these differences in the amount of
SERT protein account for a fraction—but only a tiny fraction—of the variation in the tendency to be frightened.

A similar conclusion emerged in studies of another gene, DRD4, which makes a receptor for dopamine. Scientists studied this gene because Ritalin and amphetamine stimulate behavior by releasing dopamine, which activates dopamine receptors. Researchers found that groups of people with variants of DRD4 had different average scores on novelty-seeking and impulsivity, traits expected to be influenced by dopamine.[27] Once again, the gene variants accounted for just a tiny fraction of the variation in these traits.[28]

So don't rush out to your nearest DNA testing service to have your SERT or DRD4 genes examined: They are just two examples of the thousands of gene variants that work together to influence personality,[29] sometimes in unexpected ways.[30] And even though new techniques, such as the complete sequencing of a person's DNA,[31] will eventually be used to search for variants that influence specific traits, it will still be very difficult to identify the mixture that shapes a particular personality.

But the fact that such genetic testing hasn't proved useful, at least for now, doesn't mean that you can also dismiss genetic thinking. When trying to make sense of someone, it still helps to remember that a person's specific combination of gene variants has a substantial effect on his or her personality. And we have a good idea where these many variants came from.

The Deep Roots of Our Diversity

Our view of the accumulation of so many variants in the human genome is based on Darwin's key insight that the environment keeps selecting those that increase fitness. For example, a consistent environmental factor, such as the relatively low amount of sunlight, exerted a relentless selective force on gene variants that eventually made Northern Europeans white. But Darwin also realized that environments keep changing over the course of evolution and that this led to the selection of variants that were suitable for different contingencies. Among them were those that influence personality.

To see what I mean, consider the environments that influenced the selection of the variants that control the open field behavior of mice. In dangerous territories with many predators, variants that favor caution would be selected because those who carried them would be more likely to live long enough to reproduce. But when the cats are away, the mice will play. In such safer environments, the variants that favor exploration would be selected because those who carried them might find more food and more sexual partners. Fluctuations of these alternative environments would lead to retention of both types of variants in the group's genetic repertoire. Furthermore, many of them would be kept as the species continued to evolve. This explains why some that arose in distant ancestors have been passed down to you and me.[32]

Predators are not the most important instruments of selection of heritable human personality traits. People are. They are the ones we depend on and compete with, and there are benefits and costs in the many tactics they and we use to interact. These fluctuating social environments have contributed to the selection of the wide range of gene variants that influence our personalities.[33]

In thinking about people in terms of the Big Five, it therefore helps to remember that high or low rankings on each of them have tradeoffs.[34] For example, people high in Extraversion enjoy the pleasures of intense engagement with others and the opportunities provided to those who take charge. Studies show that, like Bill Clinton, they tend to have many sexual partners, which, in a precontraceptive world, would have led to more children—Darwin's gold standard for an adaptive trait. But intense engagement comes with risks; taking charge invites jealousy and insurrection; and high excitement-seeking makes it more likely to get into accidents, engage in criminal activity, get arrested, and even get killed by rivals. So, high Extraversion is a mixed blessing.

High Agreeableness is also a mixed blessing. By promoting cooperation, it builds alliances that can pool resources for the common good and for protection against competing groups. But the downside of high Agreeableness is that it increases the chances of being taken advantage of. In contrast, disagreeable people are more likely to fight for themselves and what they believe in. Studies show that people who rank high in Agreeableness tend to earn less money, even though they are valued as team players. In contrast, those who are low in Agreeableness are more likely to rise to the top of their fields.

Great achievement is also favored by high Conscientiousness, which has the benefits of purposeful self-control and long-range planning. But high Conscientiousness has the potential downsides of oppressive perfectionism and the inability to abandon well-practiced routines in the face of changing circumstances. By always taking the long view, people high in

Conscientiousness may be less opportunistic, and this can translate into fewer sexual partners, fewer children, and less transmission of their genes. On the other hand, the children they have are more likely to enjoy the benefits of a devoted parent.

Only high Neuroticism might seem to have little to recommend it because it includes an increased likelihood of experiencing painful negative emotions. But the world can be a dangerous place, and emotions such as fear and sadness are adaptive if properly modulated. Studies show that high Neuroticism is correlated with high achievement and creativity in people whose other traits keep them from falling into the deep hole that can be dug by persistent emotional distress. Sigmund Freud, who was very high in Neuroticism, is an example.

In contrast with the assumption that high Neuroticism is always bad, most people who read books like this assume that high Openness is unreservedly good. This is because they value curiosity and are interested in new ideas. But people with low Openness are happy to exchange these pleasures for the comforts of constancy and tradition.

The fact that particular rankings on a trait have advantages and disadvantages does not, of course, mean that we consciously chose the ones we have. The reason I've pointed out their relative costs and benefits in various social environments is to help you understand why the many gene variants that influence these traits are retained in the collective human genome. Furthermore, variants that influence one trait may have been selected to balance out others. For example, it is easy to imagine how environments that favored the selection of variants for high Extraversion might have had some of their effects balanced out by the selection of variants that favored high Conscientiousness.

Such a balance of selective forces may also control the proportion of people with high or low expression of a heritable personality trait among the members of a population.[35] Consider, for example, the proportion of people with high or low Agreeableness. In a population in which almost everyone ranked high on A, the rare antisocials (with low A) would find it relatively easy to steal from their warm-hearted neighbors. These stolen resources would allow the antisocials to have more children, who would, in turn, inherit gene variants that favor low A. But as the number of antisocials increased, their high-A neighbors might band together, mount defenses to protect their resources, and turn back this growing tide. As these forces came into a stable balance over many generations, the result might be a group with a few crafty antisocials and a majority of members with a range of higher rankings on A.[36]

The Grandeur in This View of Personality Differences

The realization that human psychological diversity reflects conflicting forces of natural selection has profound implications for making sense of people. But many remain reluctant to embrace this idea because it confronts us with our primitive animal nature. Darwin himself struggled with the seemingly anti-humanistic implications of his discoveries. Nevertheless, being unable to dismiss the evidence in favor of natural selection, he eventually came to see evolution as awe-inspiring. As he explained in *Origin's* famous last sentence: "There is grandeur in this view of life ... that ... from so simple a beginning endless forms most beautiful and most wonderful have been, and are being, evolved."

To me, recognizing the role that natural selection of gene variants played in many of our personality differences is a prime example of the grandeur that comes with such understanding. And as I show in the next chapter, it has opened a way to a deeper analysis of the decades-long process by which each of us gradually develops into a unique person.

Practical Summary

In this chapter I summarized evidence that personality differences are greatly influenced by genetic makeup. If you had any doubts that a person's unique mixture of gene variants affects his characteristic ways of being, I hope you can now put them to rest.

But genes don't work alone. Culture, community, and family also shape each of us. Furthermore, it's very difficult to tease apart the genetic and environmental influences. Personalities reflect both.

In looking for reasons for the great diversity of personalities, it's also important to recognize the significance of chance. It was chance that played a part in bringing your parents together. It was chance that united your father's particular sperm cell (containing a random selection of half of his genes) with your mother's particular egg cell (containing a random selection of half of her genes) to give you your unique genetic mixture. It was chance that immersed you in the particular social and physical circumstances you grew up in and those you have lived in ever since. It's humbling, then, to recognize that so much of who you are can be traced back to these biological and environmental accidents of your birth, upbringing, and subsequent life.

Recognizing the importance of chance in shaping you should not, however, discourage you from taking responsibility for how you behave or whom you have and will become. But it can make you more understanding of yourself and of human diversity. It can also make you more restrained in your criticisms and readier to forgive.

P's Genes and Personality

1. When you've thought about the origins of P's personality in the past, have you been tempted to conclude that she was born that way? That she was raised that way? How do you see this now?

2. Does your awareness that genes influence P's personality modify your view of her? If so, how?

3. Are you familiar with P's parents, siblings, or children? Do you notice some strong similarities and differences in their personalities? Do you now find these observations more or less puzzling?

4. Does recognizing the roles of genes, life events, and chance in shaping P's personality make you more or less accepting of the aspects of her conduct that annoy you?

FOUR

Building a Personal Brain

W hen I was a fledgling psychiatrist, a colleague gave me a tip on how he gets to know a new patient. Early in the first visit, he briefly imagines the patient as a ten-year-old child. The point of this exercise is to look past someone's current troubles and picture the person as still little. Was she shy or popular? Was he a bully or a wimp?

I've found this tip useful because it immediately dials up compassion: The image of anyone as a child warms my heart. But it also creates a hunch to explore. Forming an imaginary picture of someone in grade school stimulates me to learn about the development of his personality.

When I got this tip in the 1960s, my limited knowledge of personality development was based on the ideas of Erik Erikson. A psychoanalyst who worked with children, Erikson thought we become ourselves by going through a series of well-defined stages as we progress from the extreme dependence of infancy to the responsibilities of adult life. The early stages seemed most important to him because he believed that they leave particularly enduring residues. As he explained in *Childhood and Society*:

> Every adult ... was once a child. He was once small.
> A sense of smallness forms a substratum in his mind,
> ineradicably. His triumphs will be measured against
> this smallness, his defeats will substantiate it. The
> questions as to who is bigger and who can do or not
> do this or that, and to whom—these questions fill the
> adult's inner life far beyond the necessities and the
> desirabilities which he understands and for which he
> plans.[1]

Erikson's view of personality is appealing because he reminds us of the lasting influence of childhood events. But two things are missing: genes and the brain. When Erikson wrote about the development of individual differences, he assumed that they were mainly due to upbringing and life experiences because very little was known about the influence of genetic variations. And when he described the transitions from one stage to the next, he thought of them primarily as psychological responses to a succession of challenges because very little was known about what was going on in the maturing brain.

This has changed. We now know a great deal about the way our brains develop under the guidance of our personal gene variants and our personal environments. Instead of just thinking of ourselves as solving the challenges of our youth with the brain we were born with, we have come to realize that each brain—like each face—has its own innate building plans. Furthermore, the brain's building plan was not drafted by the systematic methods of professional architects. Instead, each brain uses a scheme that would drive contractors crazy, with continuous remodeling due to changes in both genetic and environmental instructions while the project is still underway.

This continuous remodeling has a purpose. By remaining open to the interactions of our unique set of genes and environments during the more than two decades of basic construction, we each come to have a truly personal brain. Within it are the deeply ingrained components of our unique personalities that continue to guide us for the rest of our lives.

The Brain Builds Itself

The adult human brain is built of about 100 billion nerve cells (neurons), most of which were made before we were born. But not all of these neurons were created equal. As the fertilized human egg divides, it generates many types of primitive neurons, each of which is destined to play a particular role in the brain. Having been assigned their approximate fates by a process that turns on and off specific genes, the primitive neurons migrate to their designated places guided by chemical signals that they selectively respond to. When they get there, they start building connections with other neurons to form the neuronal circuits and networks that are the basis of all our behavior.

To build these connections, the neurons make branches called dendrites to receive signals and other branches called axons to send signals. Dendrites are short and studded with spines. Axons can be long enough to reach other neurons anywhere in the brain and to embrace them with clusters of little nerve endings, called boutons. Signaling between boutons of one neuron and dendrites of another occurs at structures called synapses.

A synapse is activated when a bouton releases a chemical neurotransmitter such as serotonin or dopamine onto the spine of a dendrite. The neurotransmitter travels across the synapse and binds to receptors embedded on the spine. This transmits information to the dendrite, a process called synaptic signaling or synaptic transmission.

Many types of synaptic signaling exist between neurons, governed by the dozens of different chemical neurotransmitters that are squirted from boutons onto receptors on the spines. Every neuron manufactures a particular neurotransmitter and displays a particular set of receptors. So every neuron has both a spatial address, defined by its location in a particular brain circuit, and a chemical signature defined by its neurotransmitter and receptors.

The complicated process of spatial assembly of neurons into circuits and networks is well on its way by the time a person is born. Among the circuits that operate in infancy are some in the amygdala, a brain structure that I mentioned in discussing the SERT gene. The amygdala is a hub for a complex set of circuits that integrate our emotions. Using these infantile circuits, babies experience joy, contentment, fear, anger, and the distress of separation. Neuronal controls of these emotions are gradually put in place over the next two decades, and they have major effects on the developing personality.

Circuit maturation doesn't depend only on adding new synaptic connections. While useful ones are strengthened, others are eliminated. The same selective remodeling process is also applied to the neurons themselves. Some of them grow and sprout more branches; others are destroyed by a specialized mechanism of cell death called apoptosis, which is an indispensable part of the developmental process. Much of this happens in fetal life and during the first few years after birth, but some goes on through adolescence and into adulthood.

A notable case of remodeling occurs in a group of neurons in the hypothalamus that play an essential role in the establishment of female or male patterns of sexual behavior. In the female fetus, these neurons die off as part of the developmental program that sets up female-specific sexual circuits. But in the male fetus, testosterone from the fetal testicles rescues these neurons from the apoptotic grim reaper and stimulates them to build male-specific brain circuits.[2] The timing of this effect of testosterone is crucial. If it comes too late in fetal development, the key neurons in the hypothalamus are already dead, and the brain is set on an irreversible female course. Other regulators of neuronal death may also have decisive behavioral effects, but none is as obvious as testosterone.

Brain circuits can also be modified by progressively wrapping axons with a fatty substance called myelin. Myelin acts like the insulation around an electrical cord, which facilitates the speed of conduction of electrical signals. Myelination is often a final and essential step in the genetically controlled development of a circuit.

Although this overall developmental program is at work in all of us, each of our brains is different because their structural details are influenced by thousands of gene variants in our personal genomes. There is also a little sloppiness in the assembly process, due to random variations in the movement of neurons and in the expression of critical genes. This is one reason that even the brains of identical twins are not exactly the same.[3]

Understanding the step-by-step nature of brain construction explains why it is so difficult to go back and make changes in brain circuitry and in the aspects of personality that the circuits control. Once neurons have taken up their positions, they are pretty well settled. Once they have established useful connections, those connections tend to be maintained. Although there is always some residual capacity for change, it takes a lot of work to remodel structures that are built by a developmental program that unfolds over more than two decades. Even our extraordinary human ability to learn new things may not be up to the challenge of modifying patterns that were laid down in this way. This is true not only of patterns that were strongly influenced by genes. It is equally true of those patterns that were shaped by our personal environments during phases of brain development called critical periods.

Critical Periods in Brain Development

A critical period is a window in time when certain brain circuits are open to essential environmental information. Arrival of this information shapes the circuits in a lasting way.[4] When this shaping is complete, the window is closed.

The most famous example of a critical period comes from Konrad Lorenz, who studied the behavior of baby geese. Lorenz found that each baby is primed to pay special attention to the first moving creature it sees after hatching—generally, its mother. This information is immediately imprinted in its brain, which leads it to follow its mother in those cute little trails of goslings. But if the mother goose is removed during hatching and replaced by another moving creature—such as Lorenz himself—the babies may imprint on him instead. The result is recorded in pictures of goslings trailing the bearded scientist.

Another well-known example is the development of the vocalizations of male songbirds, and it, too, involves a social interaction. In this case, the critical period of brain development is not confined to the minutes after hatching, but lasts for a few months. During this time, each juvenile male bird shapes its simple innate song by progressively matching it to the complex song of an adult male.[5] Without such instruction during this critical period, it will never be able to sing like an adult.

These critical periods in goslings and songbirds provide the opportunity to incorporate essential environmental information that is uniquely valuable to each species. For humans, a notable example is learning to speak, which develops during a critical period that lasts for more than a decade.[6] During this period, children don't only learn their native language. They also pick up the accent of the people they grow up with, especially their peers.[7] As this critical period closes, it becomes very difficult to speak like a native. This is why immigrants such as Henry Kissinger, who learned English in his teens, speak with a foreign accent. Even natives who migrate to a different region can be spotted in this way: Four decades in California have not erased the vestiges of my own linguistic imprinting in New York City.

Although researchers have studied these critical periods of brain development for many years, we still have limited information about their number and the ways they are closed. But the main message is clear: Certain brain circuits become established at particular times, and their properties tend to endure. A similar process appears to be at work in the development of many aspects of our personalities.

What Will My Child Be Like?

Although a baby is born with an immature brain, it immediately becomes a player in the world. At first, it can only cry to signal distress or coo to signal contentment. But its behavioral repertoire grows rapidly in its first few years of life as it builds new brain circuits and remodels others.

As brain development continues, parents begin wondering if their child's early patterns of behavior can provide clues about his or her mature personality. Researchers have tried to answer this question by examining children repeatedly from infancy to adulthood. Because each research group uses its own system for describing behavioral patterns, it's difficult to compare the results. Nevertheless, there is general agreement that early patterns persist in some children, whereas other children change a lot.

Evidence for some persistence of patterns comes from pioneering studies by Stella Chess and Alexander Thomas,[8] a wife-and-husband team of child psychiatrists. From their observations of babies, they identified three broad patterns of behavior, which they called temperaments. Forty percent of the babies were called "easy" because they approached new situations without difficulty, had high adaptability to change, accepted most frustration with little fuss, and were not very moody. In contrast, the 10% of the babies who were called "difficult" were much more inclined to be irritable, showed intense negative emotions, and had trouble adapting to change. Another 15%, called "slow to warm up," were initially uncomfortable in new situations but adapted after repeated contact. The remaining 35% showed a mixed picture.

Follow-ups of the children as young adults indicated that there were "only modest levels of consistency in temperament over time for a group of subjects as a whole." Their conclusion in 1986 fits well with what we know today: "Maturational factors, neurophysiologic changes, and a host of environmental influences—all these serve to produce continuity in some individuals and change in others."[9]

A series of studies led by Jerome Kagan also found evidence for both continuity and change. Kagan identified subgroups of children that he called inhibited and uninhibited, based on their willingness to engage with unfamiliar people when they were 2 and 7 years old. When he re-examined them in adolescence, he found that the majority of the children in the inhibited group remained quiet and serious, while only 15% were as lively and talkative as the average teen from the uninhibited group. Of the children in the uninhibited group,

40% maintained that style as teens, and only 5% had become subdued and quiet. As Kagan summed it up, "[A]bout one-half the adolescents retained their expectable demeanor, while only 15 percent had changed in a major way."[10]

Some behavioral continuity of members of the two groups was also observed at age 22. In brain imaging studies, the inhibited group showed significantly more activation of the amygdala when shown pictures of unfamiliar faces.[11] This sign of a stronger emotional response to new faces is reminiscent of their greater wariness of strangers as toddlers. Other researchers have also found that children retain many of their characteristics as adults.[12]

Evidence of continuity into adulthood is particularly strong for a subgroup of children who show signs of antisocial behavior in grade school. If they are sufficiently aggressive and impulsive to be singled out as having a conduct disorder before the age of 10, they tend to maintain this antisocial pattern when they have grown up.[13] In contrast, children who don't show signs of antisocial behavior until their teens are more easily reformed and have a better chance of becoming law-abiding adults.[14]

Other kinds of behavior that are prominent in childhood may change dramatically in adolescence, including some behaviors that are known to be heritable. For example, heritable childhood fears of heights, snakes, or blood frequently disappear by the time the children are in their teens.[15] How these and other waxing and waning genetic effects eventually play out also depends, in part, on interactions with the person's environment.

Gene–Environment Dialogues

Persuasive evidence of the combined effects of environment and genes comes from studies of people with antisocial personalities.[16] Everyone who has watched *The Sopranos* knows that antisocial behavior runs in families, and studies show that 10% of a community's families commit most of its crimes.[17] So you won't be surprised to learn that studies with twins show a 40% to 50% heritability of antisocial traits.[18] But in this case, family environment also has a significant effect. Furthermore, adopted children raised in antisocial families have an increased risk of developing an antisocial personality pattern,[19] even though they are genetically unrelated.

Added support for the importance of family environment comes from a study of a group of children in Dunedin, New Zealand. The researchers enrolled all of the 1,037 children born in this city from April 1972 through March 1973, assessed them at multiple intervals through the age of 26, and stored the data for subsequent analysis. This provided detailed information about child development in the entire community without preconceptions about what might show up.

One notable finding was that many of the children were abused: 8% had "severe" maltreatment, 28% had "probable" maltreatment, and only 64% had no maltreatment.[20] But this should not be taken to mean that New Zealanders are particularly nasty. In carefully controlled interviews of 8,667 American adults, 22% reported sexual abuse during childhood, 21% reported physical abuse, and 14% reported witnessing their mother being beaten; many reported all three.[21] A substantial portion also described repeated emotional abuse.

Having detected considerable child abuse in Dunedin, the researchers wondered whether it was correlated with the development of an antisocial personality pattern. To answer this question, they concentrated on boys because they are more likely than girls to develop this pattern. They found that the degree of maltreatment of the boys was, indeed, correlated with the degree of antisocial behavior. But there was considerable individual variation. Some of the severely maltreated boys developed a troublesome antisocial pattern, whereas others did not.[22] Why?

One possibility is that the boys who became antisocial had a genetic predisposition to turn out this way. For example, they might have been innately defiant or aggressive, which might have called forth more abuse. Such interactions between a child's innate tendencies and parental reactions are one reason children raised in the same family turn out to be so different,[23] and this likely played some role in the Dunedin study. But in this case, the researchers decided to get more specific by looking for a single gene variant that influenced the antisocial outcome.

To get started, they examined a plausible suspect: the MAOA gene. This gene makes monoamine oxidase-A, an enzyme that degrades serotonin, norepinephrine, and dopamine, three neurotransmitters that control brain circuits involved in emotional behaviors. Two characteristics of the MAOA gene made it seem relevant: brain levels of monoamine oxidase-A influence many types of antisocial behavior;[24] and variants of the MAOA gene's promoter control the manufacture of different amounts of the monoamine oxidase-A enzyme in the brain.[25] Furthermore, the MAOA gene happens to be located on the X chromosome, which simplifies its study in boys because they have only one copy (girls have two). In the Dunedin study, 63% of the boys had the high-MAOA variant, which makes a lot of the enzyme in the brain, and 37% had the low-MAOA variant, which makes less of it.[26]

Is having the high- or low-MAOA gene variant correlated with antisocial behavior? The researchers found that, by itself, it is not. Boys who hadn't been abused had little antisocial behavior, regardless of which variant they had. But among the abused children, there was a significant effect. Those abused children with the low-MAOA variant were more likely to become antisocial.[27]

Several subsequent studies of antisocial men support these findings.[28] So does a study of women from an American Indian tribe who had experienced childhood sexual abuse.[29] In this case, too, abuse was correlated with an antisocial pattern of behavior, and those abused women with two copies of the low-MAOA gene (one on each of their X chromosomes) had the highest rate of antisocial behavior. In contrast, those with two high-MAOA genes had the lowest rate of antisocial behavior. Furthermore, as with men, the MAOA gene didn't matter in the absence of abuse.

This doesn't mean that being born with the low-MAOA variant is bad news. Having more or less monoamine oxidase-A has multiple effects on brain functions,[30] and these may have desirable or undesirable consequences. The outcome depends on individual circumstances, other gene variants, and one's taste in personalities. The big story from the studies of childhood abuse and MAOA is more general. It illustrates the principle that genetic differences can influence the effects of childhood environments on a personality.

Enduring Effects on Gene Expression

It also works the other way: Environment can have enduring effects on the expression of particular genes that affect behavior. A good example comes from studies in Michael Meaney's laboratory of the effects of rat mothering on the personalities of their pups. The studies began by comparing the behavior of the offspring of two types of rat mothers: high-lickers who licked and groomed their pups vigorously, and low-lickers who were less enthusiastic.[31] When these offspring were tested months later, those raised by the high-lickers were less fearful and less reactive to stress than those raised by the low-lickers. Furthermore, their greater emotional stability was apparent not only in behavioral tests, such as open field activity, but also in their blood levels of glucocorticoids, stress-related hormones released from the adrenal gland.

Was the greater emotional stability of the highly licked pups caused by the maternal behavior (nurture)? Or did the high-licking mothers also have genetic differences that were transmitted to their pups via their DNA (nature)? To find the answer, pups born to high-licking mothers were swapped immediately after birth with those born to low-licking mothers, the adoption tactic that Galton had proposed to distinguish nurture from nature. The results of this cross-fostering pointed to nurture, the maternal behavior, rather than the maternal genes. High-licking foster mothers did just as good a job as high-licking biological mothers in producing stress-resistant pups, and vice versa.

Having observed this behavioral result, Meaney and his colleagues looked for differences in the brains of the two groups of pups. They found that the highly licked animals had a more active form of the gene that makes the glucocorticoid receptor (GR), a protein that responds to glucocorticoid hormones. This change, which was observed in neurons in brain circuits that control emotions, was already detectable in the pups' brains during the first week of nursing and was maintained throughout their lives.

To find out how this came about, the researchers searched for modifications in the promoter part of the GR gene, which regulates the gene's activity. It is known that promoters can be modified by a natural biochemical reaction, called an epigenetic change (from the Greek epi, which means "over" or "above"), which adds or removes a tiny methyl group at a precise point in their DNA, and that an epigenetic change may modify the promoter's effectiveness and alter the activity of the gene. The researchers discovered that the promoter of the GR gene was less methylated in the highly licked animals and that this change of their brain DNA, which was caused by their mothering, led to an increase in the manufacture of the gene's protein product, the glucocorticoid receptor.[32]

Furthermore, the behaviorally induced change in the methylation of the gene's promoter was maintained in the highly licked animals as they grew up. So, too, was the activity of the

GR gene. This suggested that the enduring epigenetic change in the DNA of these animals, and the resultant increase in glucocorticoid receptors, had shifted the settings of a brain circuit that controls the stress response. The result was a sustained effect on their personalities.[33]

The research with high-licking mothers has attracted a lot of attention because it has something for everyone. Geneticists like it because it demonstrates the importance of an environmentally induced chemical modification of a gene. Psychologists like it because it shows that behavior can affect genes as dramatically as genes can affect behavior. Neuroscientists like it because it adds to their understanding of the ways that experience can produce a sustained change in brain circuits. And all of them have been struck by the practical implications of the finding that experiences, especially those in early life,[34] can produce epigenetic modifications of DNA that have enduring effects on personality.

Such environmentally induced epigenetic changes keep accumulating as we grow up. One way we know this is from studies of identical twins. Derived from a single fertilized egg, these twins start out with identical DNA. Nevertheless, the methylation pattern of their DNA becomes progressively different as the twins grow older.[35] These epigenetic differences in the DNA of identical twins are believed to be due, in part, to the many differences in the environments the two twins grew up in. Although the functional significance of these epigenetic differences is not yet known, it is reasonable to assume that they give rise to some of the observable differences between identical twins, including differences in their personalities.[36]

Adolescent Remodeling

Although a great deal of brain development takes place in fetal life and childhood, extensive remodeling also occurs in our teens. Some of this structural remodeling is initiated by a few thousand specialized neurons in the hypothalamus that trigger the hormonal changes of puberty. These neurons make a small protein, gonadotropin-releasing hormone (GnRH), which signals the pituitary gland to activate the ovaries or testes to secrete estrogen in girls and testosterone in boys.[37] Bursts of these hormones then modify, enlarge, and activate the brain circuits for sexual behavior that were first built in the fetus.[38]

The sex hormones also do much more. By activating neurons that have receptors for estrogen or testosterone, they change the activity and settings of many other brain circuits. This gives rise to behavioral changes that are typical of adolescence, such as increased sexual interest, risk taking, impulsivity, and social awareness.[39]

But sex hormones are only one factor in the brain remodeling and behavioral changes of adolescence. Many other sex-specific changes in brain gene expression don't depend on these hormones. Both the hormone-induced and the hormone-independent processes lead to enduring modifications in brain circuits, some of which distinguish male from female brains.[40]

As in other periods of brain development, adolescence provides opportunities for genetic variations to make themselves felt. For example, some gene variants that influence cognitive abilities may not exert their full effects until the mid-teens. We know this, in part, from studies of the IQs of adopted children. These studies show that their IQs become progressively more like those of their biological parents during adolescence, as the influence of gene variants that influence cognitive abilities becomes more apparent.[41] This increasing effect of the gene variants that influence cognitive abilities was confirmed in a study of 11,000 pairs of identical or fraternal twins. The researchers found that the heritability of general cognitive abilities increased from 41% at age 7, to 55% at age 12, and to 66% at age 17.[42]

Adolescent brain remodeling is not apparent solely from the behavioral changes of the teen years. It has also been observed directly by magnetic resonance imaging (MRI) of brain structures at various stages of development.[43] The most extensively studied anatomical changes are those in the front part of the brain, especially the prefrontal cortex, which sits behind the forehead. As adolescence progresses, changes take place in the structure of regions of prefrontal cortex and their connections to brain regions such as the amygdala, which regulates emotional expression.

Changes in the connectivity and organization of brain networks during adolescence and early adulthood have not been observed just by looking at static brain structure. Functional magnetic resonance imaging (fMRI), which measures the activity of brain circuits during the performance of mental tasks, has also been used. These studies of mental activity reveal substantial changes in the functional connectivity of the brain in the progression from adolescence to adulthood.[44]

The long critical period of adolescence is also open to environmental influences. While the brain is actively rewiring, life goes on, and peers play extremely important roles in transmitting values and social skills.[45] This openness to peer influence is of particular interest to parents, educators, and clinicians, who would like to prevent the many troublesome personality patterns that start showing up at this stage of life.[46]

Closing Some Windows in the Brain and the Environment

When is brain development completed? MRI studies of individuals show that brain structure stabilizes at around age 25.[47] Although a little more myelination may continue for at least another decade,[48] changes that show up on brain scans after age 40 are generally signs of aging or disease rather than additional developmental remodeling. Furthermore, studies of the integrated activity of brain regions that is measured by functional MRI show that mature brain networks are also well established by young adulthood.[49]

This doesn't mean that the adult brain has become fixed and immutable. One of its most important functions is to keep learning and storing new information by making microscopic changes in the structure and function of synapses. Nevertheless, young adulthood marks a milestone in brain development, when we have largely built the personal instrument that will continue to guide us for the rest of our lives.

Development of personality follows a similar trend but lags behind. As anatomical changes in the brain are winding down in our third decade, changes in the Big Five are winding down too. Repeated testing shows considerable stabilization of a person's Big Five scores by age 20, significantly more stabilization by age 30, and a little more stabilization until about age 50.[50]

This progressive stabilization is not only due to the closing of windows of brain development. As Roberts and Caspi point out,[51] it is also due to the increasing constancy of the young adult's social environment. This is the environment that is populated by the friends, partners, and coworkers whom they have selected—and who have selected them.

The result of selecting a fairly constant social environment during young adulthood is that we subsequently spend most of our time with a limited cast of familiar people. These people provide stability because they keep behaving in ways that *we* have come to expect. They also elicit stability because they keep us behaving in ways that *they* have come to expect. This mutual stabilization of our social environment plays a big part in the creation and maintenance of the two overarching aspects of personality that I now turn to: character and sense of identity.

Practical Summary

Chapter 3 showed that we are shaped by the joint actions of genes and life events. This chapter described how these factors work together to gradually build an adult brain and an adult personality.

Building both the brain and personality continues actively from fetal life through early adulthood. Some aspects start stabilizing before puberty, while others are actively remodeled in adolescence and early adulthood. By age 25, the developmental sequence is largely finished.

Having completed this complex process and built our characteristic tendencies and patterns into the flesh of our personal brains, we are strongly inclined to stay pretty much the same as adulthood progresses. As a result we keep getting more and more accustomed to thinking, feeling, and acting like our familiar selves, which further strengthens the established brain circuits. Consistency is also supported as we settle into permanent relationships because the people we have become close to have also grown accustomed to our ways and expect us to stay as we are.

Nevertheless, we can always change, sometimes substantially. The most common impetus is a disruptive personal event such as marriage, divorce, parenthood, a new job, or a religious conversion, any of which may lead to new opportunities, expectations, and adaptations. Should the adaptations prove to be comfortable and rewarding, they may be incorporated into a newly stable personality. There are also developmental and degenerative processes in later life that may lead to personality changes.

Understanding the stages of personality development is useful as you look forward and back at the lives of people you know. In looking forward at the future of a young child, it's reasonable to expect that prominent tendencies will be maintained. But there will also be surprising maturational changes. In looking forward at the future of an adult, changes tend to be more limited, except in response to a major disruption.

In looking back at the origins of a person's particular tendencies and patterns, you may be tempted to point to heredity in some cases and to life events in others. But these speculations are almost impossible to verify. Instead, it is more reasonable to assume that every aspect of a personality is the result of the complex interweaving of a multitude of genetic and environmental influences that are extremely difficult to disentangle.

P's DEVELOPMENT AND CHANGE

1. Imagine P at the age of 10. What do you think she was like? Does this brief exercise change your view of her in any way?

2. Did you actually know P as a child? If so, what characteristics does she retain?

3. Was there anything unusual about P's adolescence?

4. Has P's personality changed much since age 25? If so, in what ways?

5. What personality changes, if any, have you noticed in mid-adulthood? In late adulthood?

6. Does seeing P's personality as the complex result of an ongoing dance between experiences and biology give you a more sympathetic understanding of what it's like to be in her shoes?

7. Are there aspects of P's personality that you would like to see her change? For her benefit? For yours?

8. Do you think it's possible to make these changes? If so, in what way and to what extent?

PART III

Whole Persons, Whole Lives

It matters not how strait the gate,
How charged with punishments the scroll,
I am the master of my fate:
I am the captain of my soul.

—William Ernest Henley, *Invictus*

FIVE

What's a Good Character?

When Benjamin Franklin was an old man, he revealed the secret of his fulfilling life. It was, he said, a technique that he had invented in his twenties to improve his personality.

The personality that Franklin began shaping was already standing on a strong foundation. Ever since childhood he was, according to his autobiography, "the leader among the boys."[1] But this same assertiveness cost him dearly by leading his father to withdraw him from the Boston Latin School, where he had been enrolled to prepare him for the clergy. Even though Franklin was at the top of his class and seemed destined for Harvard, then a Puritan finishing school, his father decided that he was too irreverent to be a minister and apprenticed the 12-year-old to his brother, James, a printer.[2]

Fortunately the work in the printing shop allowed Franklin to indulge his passion for reading and gave him the opportunity for an ambitious program of self-education. In studying essays from a London periodical, he learned to write so well that he was soon publishing satirical pieces in his brother's newspaper. He was also strong-willed enough to escape from his apprenticeship. At the age of 17, he ran away to Philadelphia with only a few coins in his pocket.

During the next few years, Franklin had his share of youthful adventures. But as he settled into young adulthood, he felt the need to take more charge of his life. To this end he decided to curb his passions, break some bad habits, and build up the moral part of his personality, generally called character.

The approach Franklin took to building good character began by identifying its essential ingredients. Franklin was already clear about the character traits that interested him, which he called "the moral virtues." But when he got down to making a list of them, he ran into the terminological problem that continues to bedevil contemporary discussions of personality because "different writers included more or fewer ideas under the same name." In Franklin's case, he decided "for the sake of clearness, to use rather more names, with fewer ideas annexed to each," and settled on 13 virtues, with brief explanations:

- **Temperance**—Eat not to dullness; drink not to elevation.

- **Silence**—Speak not but what may benefit others or yourself; avoid trifling conversation.

- **Order**—Let all your things have their places; let each part of your business have its time.

- **Resolution**—Resolve to perform what you ought; perform without fail what you resolve.

- **Frugality**—Make no expense but to do good to others or yourself; i.e., waste nothing.

- **Industry**—Lose no time; be always employed in something useful; cut off all unnecessary actions.

- **Sincerity**—Use no hurtful deceit; think innocently and justly, and, if you speak, speak accordingly.

- **Justice**—Wrong none by doing injuries, or omitting the benefits that are your duty.

- **Moderation**—Avoid extremes; forbear resenting injuries so much as you think they deserve.

- **Cleanliness**—Tolerate no uncleanliness in body, clothes, or habitation.

- **Tranquility**—Be not disturbed at trifles, or at accidents common or unavoidable.

- **Chastity**—Rarely use venery but for health or offspring, never to dullness, weakness, or the injury of your own or another's peace or reputation.

- **Humility**—Imitate Jesus and Socrates.

Having laid out his list, Franklin immediately got started in a methodical way. Recognizing that he could not acquire these virtues all at once, he set to work on them one at a time. Believing that "the previous acquisition of some might facilitate the acquisition of certain others," he arranged them in that particular order: "Temperance first, as it tends to procure that coolness and clearness of head, which is so necessary where constant vigilance was to be kept up, and guard maintained against the unremitting attraction of ancient habits, and the force of perpetual temptations." What Franklin particularly had in mind when starting with temperance was to stop drinking so much at pubs, which had led him astray in the past. So for the first week of his program, he concentrated on temperance. He then continued down the list, completing all 13 in a quarter of a year and then starting over again. Day by day he kept a record in a tiny book in which he "might mark, by a little black spot, every fault I found upon examination to have been committed respecting that virtue."

He found this daily record keeping both informative and rewarding. On the one hand, he was surprised to be "so much fuller of faults than I had imagined"; on the other hand, he was pleased with "the satisfaction of seeing them diminished." But despite his progress, Franklin kept returning to the program from time to time and always carried his list with him, even in old age. In assessing this lifetime of practice, he concluded, "[T]hough I never arrived at the perfection I had been so ambitious of obtaining, but fell far short of it, yet I was, by the endeavor, a better and happier man than I otherwise should have been if I had not attempted it."

Franklin had good reasons to be satisfied with the results. Within a decade of setting his self-improvement program in motion, he had built a printing and publishing business that would leave him well off. With this newfound financial security, he was able to pursue his interests in science and statesmanship, which led to brilliant achievements and worldwide fame. But even more than these trappings of success, Franklin was grateful for "that evenness of temper, and that cheerfulness in conversation" that he attributed to his devoted practice of "the joint influence of the whole mass of virtues, even in the imperfect state he was able to acquire them." So convinced was he of the value of his program that he kept toying with the possibility of publishing a self-help book called *The Art of Virtue*, to supplement what he had already explained in his autobiography.

Separating Character and Personality

Some of Benjamin Franklin's ideas about personality have a great deal in common with those I have discussed so far. He, too, recognized that people's individual differences could be thought of in terms of a set of traits. He, too, recognized that they are influenced by genes (which he called "natural inclination") and by environmental factors such as culture ("custom") and peers ("company"). And, being a lover of lists, Franklin would have been happy to organize his thoughts about his basic personality tendencies in terms of the Big Five.

Had Franklin assessed his own Big Five traits while drafting the self-improvement plan, he would have found much he was pleased with. The most obvious was his very high Extraversion, especially gregariousness, enthusiasm, and good humor. Also obvious was his self-confidence and freedom from negative emotions, signs of low Neuroticism, and his curiosity and creativity, signs of high Openness.

But Franklin wasn't particularly interested in these characteristics, which he considered part of his God-given temperament and which he took for granted. Instead, he was raised to believe that the most important part of personality was its moral aspect, which was acquired through personal effort. To Franklin, this meant that he could build his own character by working on those virtues that seemed in most need of improvement. He also believed that good character was his ticket to both productivity and happiness.

Franklin was not alone in this belief. Through the ages philosophers and religious leaders have encouraged the development of good character. What mainly distinguished Franklin's ideas from those of his predecessors was his elaborate practical method for self-improvement. Instead of simply singing the praises of a series of virtues, Franklin wrote out a personal to-do list and a step-by-step plan for upgrading one virtue at a time. Recognizing that backsliding is natural, he committed himself to repeated practice. Recognizing that some virtues, such as humility and order, were particularly hard for him to achieve, he decided to lower his standards and cut himself some slack. The result was a program that was explicit, realistic, and, as he looked back on it, seemingly effective.

Over the years, Franklin's ideas about character attracted many admirers. He also had some critics who disagreed with the list of moral virtues he chose to emphasize. But despite such disagreements, most Americans who lived in the nineteenth and early twentieth centuries shared the view that character was the most significant part of personality—and the part that could be improved through conscious effort.

Nevertheless, when scientific studies of personality were getting underway in the 1930s, the decision was made to separate the concept of character from the concept of personality. A leading proponent of this separation was Gordon Allport, whose research on categorizing personality traits I described in Chapter 1. Having been raised in a pious Midwestern Methodist family, Allport recognized that his personal values were not shared by everyone and had no place in his scientific work. As he put it:

Whenever we speak of character we are likely to imply a moral standard and make a judgment of value. This complication worries psychologists who wish to keep the actual structure and functioning of personality free from judgments of moral acceptability.... Now one may, of course, make a judgment of value concerning a personality as a whole, or concerning any part of personality: "He is a noble fellow." "She has many endearing qualities." In both cases we are saying that the person in question has traits which, when viewed by some outside social or moral standards, are desirable. The raw psychological fact is that the person's qualities are simply what they are. Some observers (and some cultures) may find them noble and endearing; others may not. For this reason—and to be consistent with our own definition—we prefer to define *character as personality evaluated;* and *personality,* if you will, as *character devaluated.*[3]

So when Allport scanned the dictionary to collect the raw material for a study of personality traits, he excluded words such as *virtuous* and *noble* that make moral judgments. Others who developed the Big Five followed his lead. Although they named some facets with moral-sounding words such as *altruism* and *modesty,* they insisted on using them in a purely descriptive way without expressing opinions about the merits of high or low scores.

The clinicians who defined the Top Ten patterns in the DSM also tried to withhold moral judgments. Trained to be open-minded about their patients' behavior, they were guided by a professional code of conduct that used functional concepts such as adaptive and maladaptive rather than moral ones such as good and bad. Their functional view recognizes that there may be advantages and disadvantages to degrees of expression of different traits and patterns, and that any of them can be adaptive in certain circumstances.

But even though this functional view appears morally neutral, it recognizes that certain patterns are worth singling out because they tend to bring grief to those who express them and to those they deal with. In fact, the negative reaction to these troublesome patterns is the main reason they are considered maladaptive. And because such negative reactions are frequently expressed as moral judgments, it should come as no surprise that features of the Top Ten are also spoken of as "character flaws" in ordinary conversation. To emphasize this point, I have listed examples in Table 5.1.

TABLE 5.1   The Top Ten as Character Flaws

Pattern	Character Flaw
Antisocial	Crooked
Avoidant	Cowardly
Borderline	Unstable
Compulsive	Rigid
Dependent	Freeloading
Histrionic	Vain
Narcissistic	Selfish
Paranoid	Untrusting
Schizoid	Aloof
Schizotypal	Bizarre

Identifying maladaptive patterns as character flaws isn't just an idiosyncratic judgment. There appears to be a widespread preference for people who are honest, courageous, emotionally stable, flexible, productive, modest, generous, trusting, sociable, and only moderately quirky—the sorts of individuals with none of the ten flaws on the list. Put simply, behaviors that we consider signs of good character may also be adaptive because most of us recognize them as being desirable, prefer to deal with those who express them, and tend to stay away from those who do not.

We make such judgments all the time. And we place heavy emphasis on character in our intuitive assessments of people. Although our minds are naturally tuned to notice all of a person's major traits, those traits that really grab our attention and dominate our thinking have a moral flavor that is linked to emotional reactions.[4]

Why is this so? Why do traits with a moral quality have such a powerful effect on us? Is this just a reflection of the cultural influences that Allport emphasized? Or does something more elemental and deep-seated about them beg for explicit attention? Might there be moral instincts that incorporate specific emotions into our assessments of people?

Moral Instincts and Moral Emotions

The idea that there are moral instincts is not new, and one of its main proponents was none other than Charles Darwin. Having recognized that instinctive social behaviors of animals evolved by natural selection, Darwin concluded that the same was also true for humans and that this process contributed to the development of our moral feelings and actions. As he put it in *The Descent of Man,* "any animal whatever, endowed with well-marked social instincts, the parental and filial affections being here included, would inevitably acquire a moral sense or conscience, as soon as its intellectual powers had become as well, or nearly as well developed, as in man."[5]

To grasp the significance of Darwin's suggestion, it helps to translate it into the language of genes. What Darwin was saying is that social and moral instincts, and the brain circuits that control them, evolved in the same way as other innate mental processes of thinking and feeling—by the natural selection of relevant gene variants—and that a reason the human genome contains many gene variants that promote social and moral instincts is that such variants contribute to fitness.

In the case of those instincts that lead people to nurture their children and to be generous to other close kin, the selective advantage is easily identified: It is the perpetuation of shared genes, the great driving force of evolution. But why would our conscience and social instincts also impel us to be generous to strangers? Given that fitness is determined by competition between individuals, shouldn't the genes that contribute to selfishness be the ones that are naturally selected? What forces would favor the selection of the genes and psychological mechanisms that restrain selfishness and promote what we call moral behavior?

A persuasive answer, which was proposed most forcefully by Robert Trivers,[6] is that moral instincts—the instincts that lead us to behave in ways that benefit other individuals or the general social order—evolved because they also benefit those who express them. Put simply, primitive forms of these instincts, such as the cautious extension of generosity to strangers, led to the selection of people who returned the favor. The mutual benefits of such reciprocal altruism—doing favors for others so that they will do favors for you—are believed to have been one driving force behind the natural selection of gene variants that contribute to morality.[7]

As with other instincts, such as our instinct to speak, scientists believe that the instinct to reciprocate evolved by modifications of brain circuits that had already become established for other reasons. In the case of the moral instincts, Frans de Waal[8] suggests that they may have had their beginnings in circuits for emotional contagion. One example de Waal gives of this primitive form of empathy is the instantaneous spread of fear from a bird that senses danger through the whole grazing flock, which immediately takes to the air. Another is the spread of crying from one infant in a newborn nursery to all the other infants in the room. In de Waal's view, such emotional contagion may have been the basis for the next type of empathy, which he calls sympathetic concern. An example is the mutual embrace of a group of infant monkeys when one of them is in distress. From these simple beginnings, new emotional brain circuits appear to have evolved[9] that immediately reward both the donor and the recipient of altruistic behavior with positive feelings.

The most obvious of these rewarding moral emotions[10] is gratitude, the feeling that wells up in us in response to kindness and inclines us to reciprocate. This warm feeling transcends any conscious ideas we have about paying someone back. Instead of reacting like robots that are programmed to give tit for tat, we appear to have evolved a tendency to feel good about returning a favor.

The same is true of compassion, which adds emotional energy to our tendency to help those in need. We don't simply make the rational calculation that someone requires assistance and that we will uphold the social order by coming to their aid. We also empathize with their pain and feel an inner sense of moral goodness as we bring them relief.

Even more unequivocally selfless is the emotion called elevation, the feeling of warmth and expansion when we simply witness or hear about acts of great kindness and compassion. If you have any doubt about the deeply ingrained nature of moral emotions, think of the tears of happiness that may come to your eyes when you observe something good happening to total strangers, tears that may flow freely not only in real life, but also while engaged in the make-believe world of the movies. Because of these properties, Jonathan Haidt has called elevation "the most prototypical moral emotion of all."[11]

But Trivers also recognized that even though these positive moral emotions provide attractive internal rewards for moral behavior, they are not sufficiently powerful to override selfishness.[12] To defend against cheaters and maintain the benefits of cooperation, we have also evolved moral circuits that are linked to negative emotions—moralistic anger, contempt, and disgust. When triggered by unfairness or by actions that seem morally repugnant, these negative emotions are usually coupled with facial and body reactions that instantly communicate disapproval and warn the violator to expect retaliation. They also generate internal feelings of indignation that may short-circuit our positive moral emotions and cause us to ostracize people who don't play by the rules.

These negative moral emotions are surprisingly easy to elicit. For example, just seeing someone cut into a line—whether we're in the line or not—may trigger moralistic anger; seeing a referee unjustly penalize our favorite football team may make us fighting mad; learning that a public figure has engaged in an immoral sexual relationship may elicit profound feelings of disgust and contempt; and even reading words that describe character flaws, such as those in Table 5.1, may arouse flickers of negative moral emotions.

Such moral condemnation is so effective because it, in turn, triggers the offender's negative emotions and makes that person feel bad. Just a simple look of disdain or disgust may instantly elicit shame, embarrassment, or guilt. As we become socialized, the desire to avoid such mental anguish may keep us from even considering actions that would make others criticize us—from wearing the wrong clothes at a party to engaging in flagrant misconduct.

The ease with which we feel these positive and negative moral emotions underscores their power. However, as with other behavioral mechanisms, there are great individual differences. Some people are strongly inclined to feel gratitude, compassion, and elevation, while others find it easier to feel disgust, anger, and contempt. Some (antisocials) cheat all the time, while others (paranoids) specialize in detecting cheaters. Some (avoidants) are especially likely to feel embarrassed, while others (schizoids) are much less sensitive to disapproving glances. But even though considerable variations exist, most of us readily experience, recognize, and respond to each of the positive and negative moral emotions. All of this fits with Darwin's idea that humans have evolved innate mental machinery that provides a biological basis for our moral behavior.

Different Cultures, Common Values

But instincts and emotions just provide the raw materials for our morality. Cultures provide the critical details. As Darwin pointed out:

> [A]fter the power of language had been acquired, and the wishes of the community could be expressed, the common opinion how each member ought to act for the public good, would naturally become in a paramount degree the guide to action ... [F]or the social instinct ... is, like any other instinct, greatly strengthened by habit, and so consequently would be obedience to the wishes and judgment of the community.[13]

Such wishes and judgments of the community vary greatly from culture to culture, and this was the reason Allport felt obliged to eliminate the concept of character from the scientific study of personality. But a group of psychologists led by Christopher Peterson and Martin Seligman have challenged that decision.[14] In a study of the major Eastern and Western religious and philosophical traditions, they found universal admiration for a large number of character strengths. The strengths that are highly valued in all cultures were combined into six categories, which they call the six core virtues:

- **Temperance**—Strengths such as self-control and prudence that protect against excess

- **Courage**—Strengths such as bravery and persistence that help accomplish goals in the face of opposition, external or internal

- **Humanity**—Strengths such as kindness and love that involve tending to and befriending others

- **Justice**—Strengths such as fairness and citizenship that contribute to community life

- **Wisdom**—Strengths such as open-mindedness and love of learning that entail the acquisition and use of knowledge

- **Transcendence**—Strengths such as awe and spirituality that forge connections to the larger universe and provide meaning

Other researchers have also recognized universally admired character strengths, and Robert Cloninger, a psychiatrist, has developed his own way of categorizing them. In his view, character has three main components, which he calls self-directedness, cooperativeness, and self-transcendence.[15] Self-directedness refers to control of the self by being purposeful, responsible, and resourceful. It overlaps with temperance and courage. Cooperativeness refers to forming mutually beneficial relationships with other people by being empathic, compassionate, and principled. It overlaps with humanity and justice.

Self-transcendence refers to awareness of our participation in the world as a whole by being spiritual, wise, and idealistic. It overlaps with wisdom and transcendence.

But recognizing the universal admiration of core virtues doesn't preclude variations in cultural emphasis. In fact, obvious differences exist in the degree to which cultures prize particular virtues. In thinking about a person's character, it is important to pay attention to the way someone expresses both universal and culture-based values.

The Power of Culture-Based Values

To study differences in culture-based values, Richard Shweder, an anthropologist, divided the moral order of each culture into three categories that resemble those Cloninger used to describe individuals. Shweder calls his categories ethics of autonomy, which resembles self-directedness; ethics of community, which resembles cooperativeness; and ethics of divinity, which resembles self-transcendence.[16]

The first of Shweder's categories, the ethics of autonomy, views each person as a free agent. Its main focus is maximizing the rights of the individual and achieving personal excellence. But the ethics of autonomy also balances the individual's right to self-fulfillment with a commitment to equal autonomy for all. It is the predominant moral view in many contemporary secular cultures.

The ethics of community turns this around by sacrificing some autonomy for the benefits of having a defined place in an organized group. It views the family and the community as the most important entities, whose moral integrity and reputation must be protected by each of its members. It also views each person primarily in terms of social roles and obligations rather than individual rights. Its main moral themes—duty, hierarchy, and interdependence—have a central place in traditional cultures.

The third category, the ethics of divinity, permeates the traditional cultures in which religion plays a major role. It views each person as a manifestation of a grand universal design that transcends individuals and provides a spiritual basis for moral behavior. In some versions, each person is seen as a responsible bearer and representative of a holy legacy rather than as a mundane practitioner of reciprocal altruism.

Breaking down a moral system into these three categories is not just an abstract exercise. It can also help us recognize how our own culture shapes our personal moral judgments. Consider, for example, something as seemingly trivial as the proper way to address your father. To most contemporary Americans, who are largely governed by the ethics of autonomy, it is acceptable to use his first name. But in the traditional Hindu society that Shweder studied in India, it is considered extremely disrespectful, a violation of both family hierarchy (community) and the sacred natural order (divinity).

The same approach can also help us understand the basis for the passionate disagreement about the morality of abortion by two groups of Americans who are each convinced that they are right. In this case, the pro-choice group belongs to a subculture that emphasizes a version of the ethics of autonomy that gives priority to the individual woman's right to protect herself from what she considers a very harmful outcome and downplays the right to life of the unborn fetus. In contrast, the pro-life group belongs to a subculture that emphasizes a version of the ethics of divinity that gives priority to the sanctity of all human souls.[17]

When considered in terms of the values of their cultures, it becomes easy to see how two people who are equally endowed with moral instincts and emotions can fervently defend such different positions. In judging the character of an individual, it is thus important to separate the person's culture-specific values from his or her rankings on those values that are universally admired. Little relationship may exist between the religious, political, and philosophical worldviews mandated by their culture and their personal rankings on temperance, courage, justice, humanity, wisdom, and transcendence.

The Character of Benjamin Franklin

To see why it's important to separate culture-specific values and universal values in judging a person's character, let's go back to Benjamin Franklin. One reason he makes a good subject is that there has been surprisingly intense disagreement about this aspect of his personality. Even though everyone recognizes Franklin's great contributions as a founding father, many critics have challenged the depth of his morality.

Much of this controversy centers on what Franklin included in his list of 13 virtues, as well as on what he left out. If you look over the list, you will see that almost all of the virtues are simply tactics for self-regulation and self-organization—the ethics of autonomy. To his fans, Franklin's practical tactics for success are worth emulating. A recent example is Stephen Covey's *Seven Habits of Highly Effective People,* which describes a stepwise plan for getting ahead that was inspired by Franklin.[18] But critics are disappointed in Franklin's focus on the practical and condemn him for neglecting the higher and more inspiring aspects of morality. He has even been accused of representing "the least praiseworthy qualities of the inhabitants of the new world: miserliness, fanatical practicality, and lack of interest in what are usually known as spiritual things …. He had a cheap and shabby soul."[19]

Walter Isaacson, who has summarized several hundred years of such polarized assessments, believes that these divergent opinions are largely culture-based, a reflection of a split in the American view of good character that was already developing in Franklin's lifetime. As Isaacson put it, "Franklin represents … the side of pragmatism versus romanticism, of practical benevolences versus moral crusading … of religious tolerance rather than evangelical faith … of social mobility rather than an established elite … of middle-class virtues rather than more ethereal noble aspirations."[20]

Franklin would probably agree about this cultural split. But he would then try to convince you that he had picked the right side. He might begin by pointing out that, instead of being purely selfish, his emphasis on self-development was also designed to help others. And instead of having a "cheap and shabby soul," he would argue that he was devoted to many high ideals, such as human rights, and had done a lot to implement them. As for spirituality, he would tell you that he valued that, too, but that he had replaced the puritanical God of his childhood with a benevolent one who "delights in the happiness of those He has created ... and delights to see me virtuous."[21] To express his belief in this benevolent God, Franklin added the following daily prayer to his table of 13 virtues:

> O powerful Goodness! bountiful Father! merciful Guide!
> Increase in me that wisdom which discovers my truest interest.
> Strengthen my resolutions to perform what wisdom dictates.
> Accept my kind offices to thy other children as the only return in my power for thy continual favors to me.

So was Franklin right to conclude that he was virtuous? Was he right to consider writing *The Art of Virtue* as a guidebook for us all? One way to assess his character is to consider how he ranked on the six universally admired categories of virtues. Focusing on them helps minimize the influence of our culture-based values.

Starting with temperance, defined as "strengths that protect against excess," Franklin readily acknowledged that he had a lot of excess to protect against. In fact, his whole self-improvement project was explicitly designed to rein himself in. So it's not an accident that his list of virtues featured efforts to control himself. And these efforts may have worked. From what we know of Franklin's life, he deserves a fairly respectable score on temperance.

Turning to courage, Franklin gets high marks. A notable example is the way he faced down vicious personal attacks from the British Crown before and during the American Revolution. And he frequently put himself in harm's way to defend causes and principles he believed in. This didn't spare him from being criticized by the more steadfast John Adams, who believed that Franklin was too willing to compromise in difficult negotiations. But to Franklin, this was a sign of shrewdness rather than cowardice.

When it comes to Justice, Franklin's score skyrockets. He was, in fact, deeply committed to fairness and good citizenship. His recognition of the value of mutual assistance was already clear by age 21, when he organized the Junto, a club of a dozen up-and-coming young men who met regularly on Friday evenings to educate and inspire each other. Franklin's interests also extended to the much larger community, which he enriched by helping to found many important institutions, from a lending library and a fire brigade to the University of Pennsylvania and the United States of America.

Moving on to humanity, Franklin deserved only a middling score for the warmth of his personal relationships. As Isaacson pointed out, "His friendships with men ... were more affable than intimate. He had a genial affection for his wife, but not enough love to prevent him from spending fifteen of the last seventeen years of their marriage an ocean away. His relationship with her was a practical one."[22] Joseph Ellis considers Franklin a master of superficial interpersonal relationships, "a man of multiple masks ... whose most sustained expressions of affection came late in life with his grandchildren."[23] But

Franklin was hardly a cold fish; closeness to others was just not his highest priority.

In wisdom, however, Franklin was at the very top: creative, curious, open-minded, eager to learn new things and to provide counsel to others. These exceptional strengths were apparent in his many practical inventions, in his famous research on electricity, and in his brilliant achievements as a diplomat and statesman. Most important, Franklin was eager to apply his broad knowledge to help others live richer lives.

But unlike Franklin's high score on wisdom, which is generally accepted, his high transcendence is not obvious to everyone. Being known for his down-to-earth practicality, many people overlook Franklin's dedication to great causes, such as religious tolerance, and to the development of the physical sciences that help us find our place in the universe. They also may fail to see that his commitment to self-improvement was not only designed to get him ahead, but was also an expression of his lofty idea that everyone can live a rewarding life if they set their mind to it. So Franklin did, in fact, find meaning in great ideals, and he certainly felt a sense of awe about the natural world. But he chose to express his transcendent feelings in practical actions rather than in flowery rhetoric.

When taken together I think that Franklin had good reason to be pleased with his character as well as his achievements. His is not a simple story of a self-made man. It is also the story of a man who was serious about his character, a man who learned to moderate his weaknesses and build on his strengths.

Proud though he was of what he had made of himself, he was also aware of his limitations and looked with tolerant amusement at those of others.

Why Character Matters

When Gordon Allport decided to "keep the actual structure and functioning of personality free from judgments of moral acceptability," he opened the way to objective assessments of individual differences in our basic traits. But sizing up people is never completely objective. When we first meet people, we don't just notice their Big Five tendencies. We also form an intuitive impression of their character.

As we get to know them better, we flesh out details of their objective and moral characteristics. But the moral ones tend to stand out because they speak most directly to our emotions, drawing us to those individuals with a mix of virtues that we find attractive and turning us away from those who do not. Although our description of a personality relies heavily on information that is contained within the Big Five and the Top Ten, we are most moved by the moral and emotional assessment of the whole package, using both universal and culture-specific criteria.

Allport recognized the importance of such moral assessments. He was just fearful that they would muddy up the rational judgments that science depends on. But this doesn't obviate our ongoing need to size people up and decide how to deal with them. Furthermore, moral characteristics also play an important part in each person's sense of identity, which I turn to in the following chapter.

Practical Summary

In this chapter I described the importance of assessing the six universally admired virtues as a third step in our system. To make the virtues easier to remember I've grouped them into categories that deal, respectively, with the person's relationship with himself, others, and the greater world:

- **Self-Directedness**—Controlling the self by being purposeful, responsible, and resourceful.

  Temperance: Self-control, Prudence

  Courage: Bravery, Persistence

- **Cooperativeness**—Forming mutually beneficial relationships with others by being empathic, compassionate, and principled.

  Humanity: Kindness, Love

  Justice: Fairness, Citizenship

- **Self-Transcendence**—Understanding and respecting one's place in the world.

  Wisdom: Open-mindedness, Love of learning

  Transcendence: Awe, Spirituality

As you think about the virtues, you will notice that each overlaps with characteristics already covered in the Big

Five and Top Ten. For example, kindness, an aspect of the virtue called humanity, is also represented in its positive form in Agreeableness, and in its negative form in the paranoid, narcissistic, and antisocial patterns. So why bother to revisit kindness yet again?

The answer is that viewing kindness as a virtue introduces an important new perspective. Whereas tendencies are evaluated on an objective scale of *high versus low*, and patterns are evaluated on a functional scale of *adaptive versus maladaptive*, virtues are evaluated on a moral scale of *good versus bad*. Having, until now, held your moral sense in check, this is the time to let it have its say by calling kindness *good* and unkindness *bad*.

Furthermore, what makes a moral evaluation so special is that it has a built-in emotional component. Linked to negative emotions such as moralistic anger, contempt, and disgust, as well as to positive emotions such as compassion, gratitude, and elevation, the scores you give someone may provoke either scorn or admiration. So this is not just a dispassionate psychological analysis. It is deeply personal. It leaves a powerful impression.

MAKING A MORAL EVALUATION

In making a moral evaluation I go through four steps:

1. I start by being aware that I am switching from a relatively objective and descriptive mode to a more subjective and judgmental one. I justify

being judgmental because I try to rely on universal standards of morality and feel that all of us should be held to them.

2. Recognizing this, I remind myself that, while trying to stick to universal standards, I cannot completely silence my cultural influences. I also pay attention to those of the person I am evaluating.

3. Being aware of these ground rules, I go on to a stepwise consideration of the person's six virtues. I start with temperance and courage, which focus on self-management; go on to his relationship with others, which includes humanity and justice; and end with his big-picture view by assessing wisdom and transcendence.

4. In making these moral judgments, I stick to rankings on specific virtues and avoid a global ranking of the person as good or bad. As with the Big Five and the Top Ten, the person may score high on some virtues and low on others, and it is the individual scores that tell me what I want to know.

Your Moral Standards

Because your assessment of others is guided by your own moral standards, this is a good time to spend a few minutes considering what they are. To clarify this, please answer the following:

1. Are the virtues included in Self-Directedness, Cooperativeness, and Self-Transcendence equally important to you? If not, please rank their importance as high, medium, or low. Then rank yourself on each of the six, using good, medium, and bad.

2. Is your standard for each virtue strict, moderate, or relaxed?

3. Do you feel strong positive emotions for people who meet or exceed your standards and strong negative emotions for those who clearly miss the mark? If so, which virtues are you most passionate about?

4. To what extent do you think your answers to these questions are culturally determined? To what extent might they be innate aspects of your personality? Might your answers to these questions strengthen or weaken your moral judgments and the passion that accompanies them?

P AND THE SIX VIRTUES

Now that you've established your moral positions, let's turn to P's.

1. Please rank P on each of the six virtues in the order of Self-Directedness, Cooperativeness, Self-Transcendence, using the scale good, medium, and bad.

2. Do you feel strong positive or negative emotions along with any of your rankings of P?

3. Have you paid enough attention to the culture-based aspects of your evaluation? For example, if you and P are on different sides of the pro-life debate, how have you dealt with this?

4. Having gone through these steps in assessing P, do you see the value of adding a moral perspective to the Big Five and the Top Ten? Do you see why all cultures place great emphasis on the moral aspect of personality, called character?

SIX

Identity: Creating a Personal Story

Until now, I've considered the aspects of personality that can be broken down into tendencies, patterns, and virtues. But to understand someone, we need to know more. Although we can piece together a revealing profile from these components, we can't complete the picture without information about the guiding principles of the person's life. To get this information, we need to shift our attention to his or her personal story.

Creating stories is one of the basic functions of the human mind.[1] It is our way of organizing sequences of experiences by inferring cause-and-effect relationships that can help us predict future events. In sizing up people, we use this process to create stories about how they got to be the way they are. Within these stories are our inferences about their motives, where they're headed, and what we can expect from them. It is our way of converting all the mental snapshots we have taken into mental movies of their lives, with flashbacks of critical episodes and projections about what will happen next. We use the same narrative process to compose stories about ourselves.

Composing stories begins in childhood, and critical events during that period have enduring effects on our developing personalities. But the stories we are mainly interested in are not simply records of objective biographical details. They are, instead, imaginative interpretations of who we are—interpretations that we begin working on seriously in our early teens. As this process unfolds in young adulthood, it gives rise to our sense of identity with which we steer the course of our lives.

This chapter is about the sense of identity, the subset of personality that psychologist Dan McAdams defines as "the personal myth you construct to define who you are."[2] Although tendencies, patterns, and virtues contribute to the creation of this personal myth, they don't tell us what it is. To grasp it, we need to learn what makes a person's life feel unified, purposeful, and meaningful, a view expressed in the form of a self-defining story.

Erik Erikson, whom you met in Chapter 4, first recognized the importance of such self-definition as an essential step in growing up.[3] In his view, the adolescent challenge of reconciling goals and interests with social opportunities and expectations is what leads us to construct an initial draft of identity. To meet this challenge, we each develop our own characteristic ways of dealing with the world,[4] along with an overall sense of who we are. We do this gradually and intuitively without much conscious thought.

Some people make this process seem easy. By their mid-teens, they have an idea about the kind of person they want to be and the path they intend to follow. This is easier to achieve in traditional societies with limited and well-defined choices. But it also happens in complex modern societies. For example, Supreme Court Justice Elena Kagan made clear her interest in becoming a judge while still in high school, and she even posed in a judicial robe for her yearbook.

Others have a harder time deciding who they are. They may find it so difficult to bring their abilities, goals, and ideals in line with social demands that they drop out of school or quit their jobs. In certain cases, this inner struggle continues well into adulthood, a condition that Erikson personally experienced and that he called an identity crisis.[5] It took him many years to settle on his identity, which he continued to work on for the rest of his life.

Paying attention to a person's sense of identity is important because it can put you in his or her shoes. Unlike the analytic understanding that comes from making a list of traits and virtues, learning about a person's view of the past and hopes for the future promotes empathic understanding. Considering the noteworthy events and circumstances in the narrative of someone's life may encourage you to identify with the struggles she encountered, the failures she experienced, and the strengths she displayed. Putting yourself in someone else's shoes may also lead you to think about who you might have become if you had been in the same situation, and this will often help you clarify your judgment of her character. To see what I mean, let's consider a famous story.

Oprah Winfrey Shapes Her Identity

Oprah Winfrey's Wikipedia page begins with a paragraph of superlatives that describes her as follows:

[A]n American television host, actress, producer,
and philanthropist, best known for her self-titled,
multi-award-winning talk show, which has become
the highest-rated program of its kind in history. She
has been ranked the richest African-American of
the twentieth century and beyond, the greatest black
philanthropist in American history, and was once the
world's only black billionaire. She is also, according to
some assessments, the most influential woman in the
world.

What makes these achievements all the more remarkable is
that they were hardly predictable from her turbulent early life.

Born in 1954 to a teenage mother from a small town in
rural Mississippi, Oprah was initially raised by her maternal
grandmother and other members of her extended family. But
this stability ended when Oprah was six. At first she went
to Milwaukee to be with her mother, and then she was sent
to Nashville to live with Vernon Winfrey, who, at the time,
believed he was her biological father. In 1963, after struggling
with Vernon's strict discipline, Oprah returned to Milwaukee.

This new environment brought sexual activity and abuse.[6]
It started when Oprah's cousin reportedly raped her when she
was nine. By her own admission, Oprah was also sexually pro-
miscuous from an early age. Her younger sister claims that,
at 13, Oprah was even selling sex to boys at her house while
her mother was at work.[7] Feeling that she couldn't control her,
Oprah's mother sent her back to Vernon.

It could have been too late. Shortly after Oprah arrived at Vernon's, in time to enroll in the first integrated class at East Nashville High School, it became apparent that she was pregnant. In February 1969, having just turned 15, Oprah delivered a baby boy.

So far, this all sounds like the familiar story of a poor child born out of wedlock who replicates her mother's struggles. But in Oprah's case the baby, who was born prematurely, died about a month later. As Vernon told her, "God has chosen to take this baby, and so I think God is giving you a second chance."[8] It was, like the pregnancy itself, one of those fateful events that can shape the course of a life. To Oprah, it meant putting everything behind her and behaving as if it had never happened.

Oprah could move on so easily because she had already rejected the possibility of settling for the role of unwed teenage mother. At 15, she had big plans for herself and wouldn't let anything stand in her way. Furthermore, the future she envisioned—to be a famous entertainer—would be based on the talents and personality traits that already made her an engaging performer as a little girl.

Those talents and traits were obvious from the time Oprah was three, when she wowed her congregation by reciting Bible stories in church. And, as she tells it, her decision to go on stage had already crystallized when she was ten, while watching Diana Ross's enthusiastic reception on *The Ed Sullivan Show*. To Oprah, who considers that the moment when her identity began to gel, the success of the glamorous African-American singer convinced her that she, too, could become a star. Even though she went through an adolescent period of wildness, she continued to believe that she was destined to be famous, and she kept her eye on that prize.

After the birth of her baby, the wildness subsided and she grabbed her second chance. This was also a time when affirmative action was beginning, and Oprah's integrated high school brought new opportunities, including classes in speech and drama that prepared her to win oratory contests. While still in high school, she also got a part-time broadcasting job at Nashville's African-American radio station. Instead of descending into the dead-end role of unwed teenage mother, the 17-year-old Oprah was envied by her classmates and already was becoming larger than life.

More success followed. Her radio performances soon led to a job at a local television station, and, at age 20, Oprah became Nashville's first black female TV personality. A few years later she was hired to anchor the evening news in Baltimore. Then, after some setbacks, the seasoned 29-year-old moved on to Chicago to build what soon became the nationally syndicated *Oprah Winfrey Show*.

While professional achievements continued, Oprah's personal life was not very satisfying. Throughout her twenties, she had stormy relationships with men who didn't stay with her. She also struggled with her weight, which had ballooned to 233 pounds when she arrived in Chicago. But instead of trying to hide her own problems, Oprah learned that she could turn some of them to her advantage.

The most famous example came in a 1985 show about childhood sexual abuse in which a tearful Oprah unexpectedly revealed that she, too, had been raped as a child. Rather than being pitied as a helpless victim, she was pleased to find herself admired as a symbol of resilience and a fearless spokesperson for the rights of women. Her obesity was also transformed from something shameful to a challenge that she could share with her viewers, many of whom had a similar problem.

The public's sympathy for Oprah's struggles stimulated her to reshape her personal myth. Instead of just aiming to be a glamorous star like Diana Ross, she became a champion of self-acceptance and recovery. Over the years, Oprah even started to think of her role as a service to a higher cause. As she herself put it, "I am the instrument of God .... My show is my ministry."[9] This spiritual aspect took many forms as her stardom increased.

Identity as a Story

Oprah's story makes good reading because she became so successful. But it also illustrates the general factors that influence the way the rest of us form mental pictures of who we are. Each case involves a constellation of traits and talents that reflect, in part, the genes we happen to have been born with. Each case involves influential life circumstances, such as gender, family, social class, nationality, culture, ethnicity, religion, and ongoing world affairs. Each case involves chance events, opportunities, and encounters that we react to and become deeply affected by. In each case the interplay of these factors is sorted and integrated to generate the characteristic ways we deal with our world. In each case these coalesce into an internal sense of principles and goals. And even though they are mainly formulated without much conscious thought, each of us sums up our version of the result in the form of a story.[10]

In Oprah's case, the story she developed is one of talent overcoming deprivation, abuse, racial prejudice, and teenage mistakes; of ambition leading to opportunities; of hard work leading to professional advancement; and of the gradual realization that her own self-acceptance can teach and inspire others. To fill in the details, she tells us that she knew from an early age that she could be a star; that even though she faltered because of mistreatment and personal failings, she didn't let this stop her; and that, in the end, she is serving God's purpose as well as her own.

Is this really Oprah's story? How much is she making up? What is she leaving out? The same questions can be asked of each of us. And the reason we find it hard to answer them is that we all have been greatly influenced by events and encounters whose impact we may be unaware of, including many that were accidental.[11] Even when we made deliberate decisions about work or relationships, they may have affected us in ways that we don't really understand. As our identity formed, important memories were unconsciously modified to conform to the internal self-image we were creating, and the past was shaped to make a more coherent story. Here is how Erikson described the development of an identity by creating a personal story:

> To be adult means, among other things, to see one's own life in continuous perspective, both in retrospect and in prospect. By accepting some definition of who

he is, usually on the basis of a function in an economy, a place in the sequence of generations, and a status in the structure of society, the adult is able to selectively reconstruct his past in such a way that, step by step, it seems to have planned him, or better, he seems to have planned it. In this sense, psychologically we do choose our parents, our family history, and the history of our kings, heroes, and gods. By making them our own, we maneuver ourselves into the inner position of proprietors, of creators.[12]

In Oprah's case, her relatives have questioned some of her selective reconstructions. For example, a cousin has challenged her memory of an extremely deprived childhood: "She's not straight with the truth. Never has been.... You should've seen the clothes and dolls and toys and little books that Aunt Hat brought home for her ... the ribbons and ruffled pinafores."[13] Members of her family have also disputed Oprah's description of childhood sexual abuse.[14] But no one would deny that she experienced hardships as a little girl who was shuttled between parents in different cities. Nor would they deny the difficulties she faced while pregnant at 14 and dealing with the premature birth and then death of her baby boy. So even though there's some uncertainty about the details, her story can still be properly told as one of recovery from adversity and as a triumph of talent, hard work, and determination. And even though what we know about Oprah's story is surely incomplete, mulling it over helps us understand her better.

Benjamin Franklin also made liberal use of selective reconstructions. As Walter Isaacson pointed out in discussing Franklin's inventions, "the most interesting thing that Franklin invented, and continually reinvented, was himself. America's first great publicist ... he carefully crafted his own persona, portrayed it in public, and polished it for posterity."[15] Nevertheless, the story Franklin told in his *Autobiography* still gives us a good idea of what he was really like.

Erikson wasn't put off by such inventiveness. Instead, he believed that inventive interpretations are essential to building a coherent identity. This is particularly important in adolescence, when we may be attracted to ideas and attitudes that differ greatly from those we were raised with and then struggle to reconcile. To bring change and continuity together, we seek out friends and environments that support what we want to become, while consciously and unconsciously inventing a story that explains this new synthesis to ourselves.[16]

The story becomes more detailed over a lifetime as we meet new challenges, and Erikson emphasized three that present themselves after we complete a first draft of identity.[17] He called the challenge of young adulthood "intimacy versus isolation," which can be met by developing close friendships and an enduring romantic relationship. He called the challenge of middle adulthood "generativity versus self-absorption," which can be met by parenting, mentorship, and altruistic contributions to the community. He called the challenge of late adulthood "integrity versus despair," which can be met by finding a way to look back at one's whole life story with understanding and satisfaction.

To Erikson, it seemed natural to think of these challenges in chronological sequence. But he also recognized that we keep working on all of them throughout our lives. Intimacy is not confined to young adulthood, generative contributions to the welfare of others may begin before middle age, and satisfaction with the integrated self we have become does not need to be postponed until we are in a nursing home. So even though it can be useful to break down a person's story into developmental chapters, we must also recognize that their contents overlap. Thoughtfully editing all parts of a story—and the identity that it represents—is necessary not only for making plans for the future, but also for adapting to the present and accepting the past.

Complicated though this process is, we are all continuously guided by our evolving sense of our own identity and by our inferences about the identity of the people we are engaged with. And we make these inferences by looking at the past and the future through stories.

Steve Jobs Tells Three Stories

We don't just create stories internally to keep us aware of who we are. We also tell them to others to project our identity. As we get to know someone, we listen to that person's stories and tell our own. Sharing stories helps us to get to know each other in ways that are not apparent from simply observing behavior.

A good example of the informativeness of personal stories comes from a commencement address by Steve Jobs. Delivered at Stanford in 2005, it described three pivotal life episodes and the lessons he learned from them.[18]

The first story Jobs told was about his own college experience. A promising student, he started at Reed College, a small liberal arts school, when he was 17. But "after six months," Jobs said, "I couldn't see the value in it. I had no idea what I wanted to do with my life and no idea how college was going to help me figure it out. And here I was spending all of the money my parents had saved their entire life. So I decided to drop out and trust that it would work out okay."

Jobs was not, however, the ordinary dropout. Having freed himself from curricular requirements, he decided to get educated on his own terms. So he stayed at Reed for another few semesters, sleeping on the floor in friends' rooms, turning in discarded Coke bottles to get money for food, and dropping in on classes that looked interesting. Among them was a course in calligraphy that he loved so much that he later insisted on including multiple typefaces in the fonts of the Macintosh computer. From this he drew two lessons: "Much of what I stumbled into by following my curiosity and intuition turned out to be priceless later on," and "You have to trust that the dots will somehow connect in your future. You have to trust in something—your gut, destiny, life, karma, whatever."

The second story jumped over the starting of Apple with Steve Wozniak when Jobs was 20, to a low point ten years later, when he was fired by John Sculley, the man he had recruited to be its CEO. Humiliated at first, Jobs went on to new greatness at Pixar and subsequently realized that "getting fired from Apple was the best thing that could have ever happened to me.

The heaviness of being successful was replaced by the lightness of being a beginner again, less sure about everything. It freed me to enter one of the most creative periods in my life." After a 12-year hiatus, Jobs returned to a faltering Apple to preside over its spectacular rebirth.

The third story was about another low point. Diagnosed with a form of pancreatic cancer, Jobs had it surgically removed in 2004. But again he saw a lesson. Instead of slowing him down, this near-death experience reaffirmed that "your time is limited, so don't waste it living someone else's life.... And, most important, have the courage to follow your heart and intuition."

These three stories told us a lot about Steve Jobs and the way he saw himself. From the age of 17, he had the confidence, resourcefulness, self-discipline, and ambition to follow his curiosity and do things his way. When faced with a crisis at 30, he relied on these qualities to bounce back. When confronted with cancer, he relied on them again.

At the close of his speech, Jobs summed up the essence of his identity, as reflected in the three stories. He said it could be described in four words: "Stay Hungry. Stay Foolish," a motto from *The Whole Earth Catalog*. The intense motivation and curiosity that this motto implies were, to Jobs, what he was all about. And he recommended this way of seeing oneself to the new Stanford graduates.

There are, however, other ways of seeing Steve Jobs. Although his own narrative is informative, learning others' stories about him can add a lot. In "The Trouble with Steve

Jobs," Peter Elkind, an editor of *Fortune,* summed up some of those stories.[19]

Among the troublesome features Elkind identified, many can be attributed to Jobs's low Agreeableness, which is not rare among top business leaders. From what Elkind learned, Jobs "oozes smug superiority" (arrogance), "periodically reduces subordinates to tears" (heartlessness), "fires employees in angry tantrums" (combativeness), and "is notoriously secretive" (suspiciousness and deception). Elkind also found signs of all three patterns of low Agreeableness: narcissistic, as revealed by smugness and an insistence on making his own rules; paranoid, as revealed by a level of secretiveness that even his Silicon Valley colleagues consider extreme; and antisocial, as suggested by reports that "he parks his Mercedes in handicapped spaces" and that he condoned backdating of stock options.

Another troublesome feature that Elkind identified is perfectionism, the dark side of Jobs's exceptional competence. This dark side led John Sculley to call him "a zealot, his vision so pure that he couldn't accommodate that vision to the imperfections of the world" and to fire him in 1985. This dark side of his perfectionism may also have interacted with his low Agreeableness and led him to call subordinates "shitheads" and "bozos" if they didn't meet his exceptional standards.

But to Jobs, the troublesome characteristics I've mentioned might just be inconvenient by-products of staying hungry and foolish. If you asked him why he wasn't nicer to people, he might have said that it would have gotten in the way of the true excellence he was striving for. If you asked him why he didn't stop being such a control freak, he might have explained that it's all too easy to slide into mediocrity and that he just wasn't willing to lower his standards. Then he might have gone on to tell you that the ultimate justification for this way of being was apparent not only in the beauty and elegance of his products, but also in their social value and commercial success.

When viewed in this way, it is reasonable to conclude that much of Jobs's personal myth was truly represented in his three stories. There are, of course, other important stories we might want to consider, including those about his adoption and search for his biological parents; his youthful immersion in Buddhism and experimentation with LSD; the way he dealt with the birth of his first child out of wedlock when he was 23; and his relationship with his wife, Laurene, and their children—and much has been made of these stories by people who knew him well. But work appears to have dominated Jobs's life, and "Stay Hungry. Stay Foolish" was his way of explaining his approach to it.

Guiding principles that can be summarized so succinctly are not unusual. Dan McAdams, who has devoted his career to studying people's life stories,[20] finds that such principles become increasingly coherent as we settle into middle adulthood. Although some flexibility remains, to allow for adaptation to changing circumstances, inconsistencies tend to be reconciled as our stories mature. The essence of our personal myths can then be enunciated in a few simple phrases that we tell ourselves—and others—about who we are.

Practical Summary

In this chapter I illustrated the importance of including a person's self-defining story, called his identity, in a description of his personality. To assemble this story, I use everything I've learned about him and organize the information in four categories:

1. **General characteristics.** I start with the significance he attaches to characteristics such as his physical appearance, race, gender, family structure, social class, financial status, ethnicity, religion, education, and occupation. Which of these does he consider to be most important in defining who he is? Is he proud of some and dissatisfied with others?

2. **Personal influences.** I then try to identify the people he turns to for advice, encouragement, and criticism. Is it his spouse? His boss? A friend? A clergyman? A therapist? Is there a book, a song, or a quote that inspires him? Does he have a personal motto? A role model?

3. **Life story.** Having identified his most important general characteristics and personal influences, I gather up whatever stories he told me about how he got to be who he is and where he is headed. In the process I learn about privileges and handicaps, successes and failures, friends and enemies, good luck and bad luck. Does he view his earlier life with pleasure or regret? Is he content with his

present situation? Is he hopeful or pessimistic about his future? Although I am mainly interested in what I learned from him, I also consider material from other sources.

4. **Coherence and adaptability**. As the picture takes shape I try to decide how well it fits together and how much inconsistency and internal conflict I can detect. I also take note of his ability to adapt to major opportunities and setbacks.

When I'm done mulling this over, I pick out the highlights from each category and put them together as either a list or a paragraph. In doing this, I'm very selective. My aim is to assemble a readily accessible picture that tells me, at a glance, where he thinks he came from, who he thinks he is, and where he thinks he's headed. This is the picture of his identity I store in my mind.

P's IDENTITY

Using the process I've just described, please select the highlights you plan to emphasize in describing P's identity. To think this over, you may wish to consider the following questions:

1. **General characteristics**
   - How does P feel about her physical appearance and gender?
   - What about her race and social class?

- Does she believe that her ethnicity and religion play defining roles?

- Is she satisfied with her education and occupation?

- Which of these characteristics seem particularly important to her?

- Is she proud of some and dissatisfied with others?

- Is she preoccupied with any of them?

2. **Personal influences**

- How does P feel about her parents and siblings?

- Does she identify with one or more of them?

- Does she feel she had important mentors at particular points in her life?

- Does someone stand out as a positive role model, and why?

- Does she have negative role models or people who haunt her?

- Does she have close friends?

- Are there enemies or competitors she fears or hates?

- To whom does she turn for help or advice?

- Does she try to model herself after an important public or historical figure?

- Are there books, songs, or mottos that have helped her define her own worldview?

3. **Life story**

- If you asked P to sum up her life, what do you think she would say?

- Has she ever spontaneously volunteered this information?

- How much have you discussed this with her?

- Does she seem to have a clear way of thinking about the ways that the people and circumstances in her life supported or stood in the way of her interests, abilities, and opportunities?

- Does she mention privileges and handicaps, successes and failures, good luck and bad luck? Which does she emphasize?

- How does the life story she projects compare with the one you would assign to her on the basis of other information you have? Is there a big discrepancy?

4. **Coherence and stability**

- Has P given you any information about the way her identity took shape?

- Were there clear indications of an emerging identity in adolescence? Did it come together in early adulthood? Is it still open to modifications?

- How satisfied is she with her identity?

- Does she see her life as a coherent whole moving gracefully from her past to her future?

- Does she have persistent regrets about mistakes?
- Does she understand the role luck played in her life? Does she emphasize luck too much or too little?
- Does she have the ability to make peace with setbacks and chart a new course?
- Does she wish she were a different person?
- Does she believe she can change if she needs to?

As you can see from this series of questions, it's much more difficult to sum up P's identity than to assess her tendencies, patterns, and virtues. But the great value of doing it will become apparent as you write out your description. Whereas tendencies, patterns, and virtues can tell you a lot, their roles in P's personality only become clear when combined with her identity, as you will see in the following chapter.

SEVEN

Putting It All Together

I opened this book with a phrase I find both simple and profound:

Every man is, in certain respects
(a) like all other men,
(b) like some other men,
(c) like no other man.

Written in 1953 by Clyde Kluckhohn, an anthropologist, and Henry Murray, a pioneering personality researcher[1] at a time when *man* was a widely used synonym for *person,* it continues to remind me that understanding someone comes by focusing not only on his or her differences from others, but also on what we share.

Although this may seem obvious, our shared humanity is the first thing we need to acknowledge when trying to make sense of a person. Each of us has a human genome and a human brain. Each of us was once a small child. Each of us was raised in a complex culture and faces similar life challenges. Before considering what distinguishes each of us, it is important to pause for a moment to explicitly remember how much we have in common.

Having consciously recognized this sameness, we are ready to consider the person's notable differences from the many other people we know. In this book, I've described several ways of thinking about these differences. Here I will show you the value of putting them together by going back to the examples I used at the start: Bill Clinton and Barack Obama.

I chose these two familiar people because you have probably seen them in action on many different occasions. Even if you haven't given them much conscious thought, you've probably formed intuitive pictures of what they are like. It is the observations that gave rise to these intuitions that are the raw materials for more methodical appraisals of their personalities.[2]

A Methodical Appraisal

A good way to begin a methodical appraisal is with the Big Five and its facets. Although we form an initial impression of each of these tendencies when we meet someone new, we refine that impression as we see how the person behaves in various situations. So the Big Five scores we give people as we get to know them are really rough averages of many observations. Mentally computing a person's average score for each tendency also provides the opportunity to take note of interactions with other tendencies that may give rise to characteristic behavioral nuances.[3]

In applying the Big Five, I start with Extraversion because it is usually easy to assess.[4] I then continue to Agreeableness, Conscientiousness, Neuroticism, and Openness. I find it helpful to stick to this fixed order when making my initial survey, but I jump around freely when I make mental revisions. As I've gained experience in sizing up people, I still find that this simple tool helps me notice aspects of a personality that I might have overlooked.

Having considered the Big Five and its facets, the next step is to focus on what is particularly prominent. For example, Bill Clinton's exceptionally high Extraversion would top his list. His other notable tendencies, which I considered in Chapter 1, are relatively low Conscientiousness and an Agreeableness score that is lower than it may initially seem. In contrast, Barack Obama's Conscientiousness is much higher than Clinton's, and his Extraversion is conspicuously lower. Obama's Neuroticism score is also notably low—so low that his advisers must sometimes prod him to express negative emotions.

After I have identified a person's most prominent tendencies, I look for patterns. To get started, I look for evidence of four potentially troublesome ways of thinking about oneself: "I'm special," "I'm right," "I'm vulnerable," and "I'm detached." If I find a fit, I compare the characteristics of the person's pattern with those of the Top Ten. For "I'm special," I consider antisocial, histrionic, or narcissistic; for "I'm right," I consider paranoid or compulsive; for "I'm vulnerable," I consider avoidant, borderline, or dependent; and for "I'm detached," I consider schizoid or schizotypal.

Although these ten patterns were initially identified because extreme forms are maladaptive, mild forms are common and are worth looking for in everyone. Even though none of them can describe someone perfectly, these patterns are points of comparison that can help you clarify what you actually observe. If you don't see signs of any of them, that, too, is informative.

Again, Clinton and Obama are good examples. In Clinton's case, he clearly thinks of himself as special. And even though he has good reason to be proud of his exceptional talents, his intense desire to be admired brings to mind the narcissistic pattern. This need to be surrounded by enthusiastic fans is common in people who become leaders and fuels their ambition and accomplishments. Convinced of their superiority, they project confidence and, like Clinton, often bounce back from major setbacks.

Clinton also exemplifies some of the troublesome aspects of this pattern. One is a feeling of being entitled to take advantage of others. A second is a sense of invulnerability that may impair judgment. It is this combination of entitlement and invulnerability that allowed Clinton to engage in a prolonged and poorly disguised sexual relationship with a White House intern while being investigated for earlier sexual improprieties. If you wondered how he could have put himself at great risk for so little reward, and how he was so careless in covering his tracks, it may help to remember that such behavior is not rare among invulnerable narcissists. So Clinton's version of the narcissistic pattern, which has served him well in many ways, has an obvious downside.

Obama's personality is very different than Clinton's, and it doesn't bring to mind any of the Top Ten. Although Obama is every bit as special as Clinton, he doesn't share his sense of entitlement or invulnerability. Although he isn't very gregarious and enjoys his privacy, he is hardly detached. Although he has clear goals and strong opinions, he is eager to consider alternatives, offers reasonable justifications for his controversial decisions, and isn't blindly certain that he's right. And his imperturbability and low Neuroticism make him the opposite of vulnerable. In Obama's case, then, I see no signs of patterns that can be considered maladaptive.

Noticing patterns doesn't only shift attention from the description of traits to an assessment of their adaptive value. It also sets the stage for another type of judgment using criteria that are explicitly moral. Unlike adaptive criteria, which are based on observations about what works for the person, moral criteria are influenced by our instincts about good and bad. What makes moral judgments so compelling is that they are powered by positive emotions such as compassion and by negative emotions such as contempt. In deciding what you think about someone, your observations come into sharp focus when viewed through the lens of morality.

But it's important to remember that this moral lens is also shaped by the culture and subculture we belong to, and such cultural differences play a big part in the divergent moral judgments of Barack Obama and Bill Clinton. Members of some political subcultures consider one or both to be morally inspiring, while members of others look down at Obama for being morally aloof and for having no backbone, and sneer at Clinton for his flagrant misconduct and for being "slick Willie."

Such moral opinions are generally experienced as gut reactions to specific characteristics that turn us on or off. But to make sense of someone, we need to think over our gut reactions by systematically evaluating the person's character strengths and weaknesses. A good way to start is to consider how the person measures up on the three domains of character: self-directedness, cooperativeness, and self-transcendence. We can then flesh out the picture by examining the way the person expresses each of the six core virtues: temperance, courage, humanity, justice, wisdom, and transcendence.

In making a conscious moral assessment, I believe that it is also necessary to take note of the degree to which we are relying on universal as opposed to culture-based standards. I find this helpful in judging Clinton's and Obama's character strengths and weaknesses, and I invite you to now make your own moral evaluations of them with this in mind. Awareness of my cultural biases also inclines me to be more open-minded when I turn my attention to their life stories.

Thinking about life stories opens the floodgates to details that were ignored in the relatively abstract survey of tendencies, patterns, and virtues. It is in the context of people's stories that we can include the many features that contribute to their uniqueness. These range from physical characteristics such as gender and appearance; to socioeconomic, ethnic, religious, and cultural factors; to family structure and educational opportunities; to strokes of good or bad luck—a vast amount of information that might overwhelm us if we hadn't already built a tentative picture of someone's personality. Having first formulated a working idea in the ways I've already described, we can try to understand how all these other factors contributed to the person's sense of identity.

Here again, Obama and Clinton are good examples because they have told us many stories in their autobiographies. In them, they have explained their ways of thinking while dealing with a series of challenges, opportunities, and lucky breaks. So reading their books will give you plenty of clues to their personal myths. But the reading will be more meaningful if you begin with the tentative picture you've already developed by building a profile of tendencies, patterns, and moral characteristics.

Obama's and Clinton's stories have a lot in common. Both were the first children of adventurous and ambitious mothers. Both had little or no contact with their biological fathers: Clinton's father died in an auto accident before he was born, and Obama's left when he was two and later also died in an auto accident. Both spent part of their childhoods living with their grandparents. Both had stepfathers. Both were elected president in their mid-forties.

But the identities they developed and the paths they took were very different. Clinton had already opted for a life in politics by the middle of high school. In his autobiography, he says, "Sometime in my sixteenth year I decided I wanted to be in public life as an elected official ... I knew I could be great in public service."[5] At 17, he visited the White House as an Arkansas delegate to a convention that introduced high school students to the federal government, and he dashed to the head of the line to get his picture taken with President Kennedy. While in college at Georgetown, he interned with Arkansas Senator Fulbright to get a foothold in politics. A year after graduating from Yale Law School, at the age of 28, he ran for Congress. By age 32, he was Governor of Arkansas.

Obama took longer to figure out who he was. The biggest issue was that he was biracial, the son of a Kenyan father whom he described as "black as pitch" and a Kansan mother "white as milk."[6] Raised by white grandparents, Obama said that his obvious African origins made him feel different than his classmates at the elite Punahou School in Hawaii, and he spent much of his young adulthood "trying to raise myself to be a black man in America."[7] After graduating from Columbia and experimenting with life in New York City, he moved to the South Side of Chicago to immerse himself in African-American culture, work for social justice as a community organizer, and build his identity.

Having clarified his goals in Chicago, Obama was ready to reach for them by attending Harvard Law School. He was also ready to reach still higher, and he successfully campaigned for the editorship of the *Harvard Law Review.* This earned him national attention and a book contract for *Dreams from My Father,* which became a foundation for his political career. The rest is history. Blessed with intellect, education, the ability to inspire both black and white voters, and what Liza Mundy thoughtfully summed up as "a series of fortunate events,"[8] Obama jumped to the top. At age 47, he became President of the United States.

The differences in Obama's and Clinton's stories are reflected in their goals and the identities they developed. Clinton's major goal, which was already apparent in his teens, was to use his persuasiveness to become a political leader. More interested in being admired than in specific policies, he easily switched positions on important issues, which brought him both success and condemnation. Having survived a major scandal, he later stepped back into the limelight to enjoy the popularity that he so avidly seeks.

Obama, in contrast, is more interested in changing the world than in the enthusiastic approval of the crowd. His guiding theme, which has obvious personal relevance, is to bring all people together. As he tells us in *The Audacity of Hope,* "we are becoming more, not less, alike.... Identities are scrambling, and then cohering in new ways. Beliefs keep slipping through the noose of predictability. Facile expectations and simple explanations are being constantly upended."[9] Truly committed to this vision, Obama took his election in 2008 as further evidence of the growing trend toward social harmony and universal brotherhood. He retains this vision even though his efforts to implement it have been constrained by fervent political opponents.

Few stories are as eventful as Clinton's and Obama's. But the basic ingredients of everyone's stories are the same. Traits, talents, values, circumstances, and luck contribute to all our stories, and we can identify their roles in an overall picture of each personality. To put together this big picture, I find it useful to follow the steps discussed in this chapter:

1. Remember our common humanity and the way personalities develop.
2. Make a Big Five profile and notice what stands out.
3. Look for potentially troublesome patterns.

4. Make a moral assessment using universal and cultural standards.

5. Listen to the person's story and relate it to what you observe.

6. Integrate the highlights into a picture that emphasizes what you consider most important.

By following these steps, I bring order to my observations. Having built this overall view, I can then take note of those situations in which the person deviates from their general way of being, and I can enrich the picture by incorporating information about the person's inconsistencies. The result is a living picture of the person that amalgamates the pieces I've added in this stepwise manner, and that, once formed, I come to see as a whole. Although I can always deconstruct the picture to add some critical new bit of information, I access it in my mind as a revised intuition that, while still instantly accessible, has now been informed by systematic conscious thought.

Understanding and Change

Systematically building a picture of a personality doesn't just help you understand a person; it also helps you think more clearly about one of the biggest questions you may have: Can this person change?

The answer to that question depends on the characteristics you are concerned with. Big Five tendencies tend to stabilize in young adulthood and become even more stable by middle age,[10] as do personality patterns. So when you've got a clear picture of someone's tendencies and patterns, it's best to assume that what you see is what you will continue to get.

But values, which are strongly influenced by culture, can sometimes change a lot. This happens less frequently in stable traditional cultures than in more open ones, such as contemporary America, which encourage personal experimentation. Although much of this experimentation goes on in adolescence and early adulthood, some people raised in a subculture that emphasizes autonomy may later be drawn to one that emphasizes community and divinity and may even be "born again." Others raised in a religious subculture may abandon it in favor of a secular one.

Stories, and the sense of identity that they express, can also be modified. Here important life circumstances can be very influential. Marriage can have a significant effect, as can divorce. The birth of a child can be an important turning point, as can a child leaving home. Getting a great job can be transformative, as can losing it. In each case, the major event alters environmental factors that stabilize a personality, and this provides an opportunity to reconsider who we are and where we're headed.[11] Psychotherapy may also stimulate the revision of a personal narrative, and its success appears to depend on it.[12] So if you pay attention to a person over long periods of time, you should be prepared to see some changes in his or her personal myth.

But the changes you are probably most interested in are more immediate and personal. They are the changes you may be hoping for in someone you're close to. They are the changes that would transform his or her day-to-day behavior in ways that would please you.

If you have been hoping for such changes, it may help to rethink this in light of what you've learned from this book.

What do you now believe is bothering you about this person? How does it fit into the overall personality picture you've built? Is it situational? Is it cultural? Is it a response to specific environmental factors? Are you doing something to bring it out?

These questions are not designed to show you how to change this person. They are, instead, designed to clarify your understanding of the course he or she is on. They are designed to highlight the characteristics you admire and those you do not. They are designed to help you see this person in the context of those others you have also come to understand.

What may, however, change in the process is the way you choose to relate to this person. Although moment-to-moment relationships with people are always ad lib, it can be helpful to first figure out what you think of someone and how you intend to deal with him or her. Spontaneity is essential, but preparation helps. As Dwight D. Eisenhower put it, "In preparing for battle, I have always found that plans are useless, but planning is indispensable."

But in the end, the greatest value of making sense of people transcends practicality. It is the pleasure we get from understanding their differences from others, as well as their ultimate sameness. It is the pleasure we get from more fully appreciating the humanity of those with whom we share our lives.

Practical Summary

In this chapter I used the examples of Bill Clinton and Barack Obama to illustrate the value of combining all four of the perspectives I've described in this book. Now I'd like you to do the same for P.

Summing Up P

To build this picture of P's personality, please think over what you've learned about her:

1. **Tendencies**. Are any of P's rankings on the Big Five considerably higher or lower than average? If so, which stand out? Is P aware of these differences, and does she feel pride, distress, or indifference about them? Do you want to emphasize any of them in your overall description?

2. **Patterns**. Does P display any patterns of behavior that you or others find troublesome? If so, do they resemble one or more of the Top Ten? Does she, too, find them troublesome? Are any of them sufficiently troublesome to take a leading place in your overall description?

3. **Virtues**. How does P rank on the six virtues? Is she unusually high or low on any of them? Do you have strong positive or negative feelings about any of her rankings, and do you want to emphasize them in your description? Are your strong feelings based on universal or culture-based criteria?

4. **Identity**. How much has P told you about her personal story? Do you believe she has a clear view of who she is, what she stands for, and where she is headed? How much have you learned from others? Do you see her the way you think she sees herself? How would you sum up the major forces and pivotal events in her life? If you had to assign a motto to her, what would it be?

After you've considered these questions, put the highlights together in a way that portrays what P is typically like. To do this, you may find it easiest to just enumerate her prominent tendencies, patterns, and virtues and end with what you've already written about her identity at the end of Chapter 6. You may, however, find it more interesting to intermix highlights from all four perspectives into a multi-dimensional picture.

Finally, having prepared this description of P's personality, please compare it with the one you wrote down at the start of this book by answering the following questions:

1. Are the main points still the same? Or have you added some new observations that you had overlooked?

2. Did the system help you organize your observations by providing you with an orderly process and a well-defined vocabulary?

3. Did using the system increase your confidence in your assessment?

4. Did it help you decide which aspects of her personality you like and which you don't?

5. Did it make you more or less sympathetic to P?

6. Did you identify changes you might want to make in the way you relate to her?

7. Do you think those changes might have a favorable effect on the way P relates to you?

I hope that your answers to these questions indicate that this system has been useful. Should this be the case, here are some benefits of continuing to apply it in your everyday life:

1. **The system augments the assessments you're already making.** Even though you've already had a lot of experience in sizing up personalities, the system will help you collect and organize information into a more complete picture.

2. **It starts working with a first encounter.** To use the system to full advantage, you need to learn a lot about a person. But it also starts paying off in a casual social encounter by immediately suggesting things to look for. And it may be particularly valuable in a planned first meeting, such as a blind date, by helping you notice characteristics to think over later.

3. **It becomes more efficient with disciplined practice.** To get better at using this system, it is necessary to keep making detailed assessments like the ones you've made of P. At first this may seem tedious. But should you continue to do this in a disciplined way, it will become progressively easier to pick out the highlights and come up with an overall picture.

4. **It makes you aware of your biases.** Your assessments of others are always made through the lens of your own personality. Gaining experience with this system will increase your awareness

of your attitudes toward particular personality differences and encourage a more objective point of view.

5. **It modulates your emotional judgments.** In making an assessment, it helps to be objective and unbiased. But as you collect the data, you will naturally activate moral emotions that reflect your values. Using the system will help you recognize these emotional judgments, modulate their positive or negative spin, and add some insight about your preferences.

6. **It may guide a thoughtful plan of action.** Using the system will not only clarify what to expect from a person. It may also help you figure out ways to build up your positive relationships and manage or avoid your negative ones.

Final Words

When you started reading this book, you may have worried that breaking down a personality into its parts would take the life out of it and make you miss the forest for the trees. As you finish this book, I hope I have persuaded you that mastering this system will, instead, help you put together a more vivid overall picture. Furthermore, should you stick with it, I believe you will find that the process becomes increasingly effortless and natural. Should that be the case, I hope you will also find yourself enjoying a richer understanding of everyone you meet.

Endnotes

Preface

1. http://georgewbush-whitehouse.archives.gov/news/releases/2001/06/20010618.html

2. http://www.hughhewitt.com/president-george-w-bush-and-decision-points/

3. Bush's portrait of Putin is at www.flickr.com/photos/georgewbushcenter/sets/72157643401817945

4. http://www.today.com/news/bush-putin-dissed-our-pet-said-his-dog-was-bigger-2D79484676

Introduction

1. Bargh and Chartrand (1999), and Bargh and Williams (2006) sum up evidence that we engage in most of our social interactions without conscious awareness. Gigerenzer (2007) and Gigernezer and Goldstein (1996) emphasize the advantages of being guided by unconscious gut feelings. Gladwell (2005) calls this a "blink" and "thinking without thinking."

2. Wilson (2002) makes a lucid comparison of the conscious and unconscious mental processes we use in sizing up people. He views the unconscious processes as rapid pattern detectors that have the advantage of great speed

but are more prone to errors than the slower conscious processes that rethink these immediate impressions. Whereas the unconscious processes are concerned with moment-to-moment assessments and are automatic, unintentional, and effortless, the conscious ones take more time; are controllable, intentional, and effortful; and can ultimately be very useful. Kahneman (2011) calls the rapid automatic thought process System 1, or "thinking fast," and the effortful one System 2, or "thinking slow." Epstein, et al. (1996), have described individual differences in intuitive and analytical thinking styles, and Frith and Frith (2008) have examined differences between implicit (unconscious) and explicit (conscious) processes in social cognition.

Chapter 1

1. Klein (2002).

2. Allport and Odbert (1936).

3. This dictionary-based investigation of personality traits is called the lexical approach. Its history is summed up by John, et al. (1988), and Digman (1990).

4. Allport (1961), p. 355.

5. Craik, et al. (1993), review the early history of personality research. John and Robbins (1993) and Nicholson (2003) emphasize Allport's contributions, including his interest in both building blocks of personality (traits) and the uniqueness of each whole person.

6. Long before Allport and Odbert published their findings, Galton had made a preliminary survey of "the most conspicuous aspects of the character [his word for what

we now call personality] by counting in an appropriate dictionary." In *Measurement of Character*, an article he published in 1884, he estimated that the dictionary "contained fully one thousand words expressive of character, each with a separate shade of meaning, while each shares a large part of its meaning with some of the rest."

7. Goldberg (1990, 1992, 1993).

8. Denissen and Penke (2008) believe each of the Big Five reflects the activity of a brain system that controls a particular social or general motivation. In their view, Big Five scores "reflect stable individual differences in their motivational reactions to circumscribed classes of environmental stimuli. Specifically, extraversion was conceptualized as individual differences in the activation of the reward system in social situations, agreeableness as differences in the motivation to cooperate (vs. acting selfishly) in resource conflicts, conscientiousness as differences in the tenacity of goal pursuit under distracting circumstances, neuroticism as differences in the activation of the punishment system when faced with cues of social exclusion, and openness for experience as differences in the activation of the reward system when engaging in cognitive activity."

9. Mischel (2004) has emphasized the fact that there are consistent individual differences in the expression of a trait in specific situations, which he calls "*if* ... *then* ... situation-behavior relationships."

10. Funder (1995, 2006) has demonstrated the value of averaging together our behavioral observations to form judgments about a person's relative rankings on each of the Big Five. Mischel (2004) has acknowledged

that this averaging process "has proven to be of much value, especially for the description of broad individual differences on trait ratings of what individuals 'are like on the whole.'" But he points out that a great deal can also be learned by observing the person's distinctive pattern of "*if ... then ...* behavior" in specific situations. See also Mischel and Shoda (1998) and Kammrath, et al. (2005).

11. In making assessments, you will probably find a great deal of variation among the men and the women you know, with no obvious gender differences in their profiles. Although Costa, et al. (2001), and Schmitt, et al. (2008), detected gender differences in Big Five studies in dozens of cultures (women, on average, tended to score a little higher on N, E, A, and C), there was a great deal of overlap. The magnitude of the gender differences varied from culture to culture. Surprisingly, "sex differences in personality traits are larger in prosperous, healthy, and egalitarian cultures in which women have more opportunities equal with men" (Schmitt, et al. [2008]).

12. McCrae and Costa (2003) reviewed the history of the NEO PI-R, the test they designed to be administered and interpreted by professionals (Costa and McCrae, 1992) that is widely used by clinicians and in research on personality. The test has been translated into many languages and found to be useful in many cultures (D. P. Schmitt, et al. [2007]; McCrae and Costa [1997]).

13. Questionnaires made up of phrases rather than adjectives are not new. Hans Eysenck (1965), a British psychologist, used them in his pioneering studies of personality traits. So did Katherine Briggs and her daughter, Isabel Briggs Myers, who created a widely used personality test called the Myers–Briggs Type Indicator (MBTI) (Myers, 1980),

which is the foundation of David Keirsey's (1998) popular book on personality. One reason academic psychologists prefer the NEO PI-R is that it assesses all of the Big Five, whereas the MBTI leaves out Neuroticism (McCrae and Costa, 1989; McDonald, et al., 1994).

14. Buchanan, et al. (2005); Goldberg, et al. (2006).

15. Johnson (2005). Internet questionnaires have also been studied by Gosling, et al. (2004).

16. Hitchens (1999) provides an extreme example.

17. Obama (1995, 2006).

18. Klein (2010).

19. Dowd (2010).

Chapter 2

1. American Psychiatric Association (2004). The descriptions are mainly from the fourth edition (DSM-IV). They are very similar to those in the fifth edition published in 2013.

2. I have used the term Compulsive, which is what this pattern was called in the third edition of DSM (DSM-III), rather than Obsessive-Compulsive, which is what it is called in DSM-IV, to avoid confusion with the mental disorder called Obsessive-Compulsive Disorder (OCD). OCD refers to recurrent obsessions or compulsions that cause marked distress or significant impairment and that the person recognizes to be excessive or unreasonable. In contrast, people with a compulsive personality pattern are proud of their commitment to orderliness, perfection, and control. Although some people have both OCD and a prominent compulsive personality

pattern, these are distinct entities without a lot of overlap (Mataix-Cols [2001]; Miguel, et al. [2005]; Samuels, et al. [2000]).

3. The American Psychiatric Association published the fifth edition of the *Diagnostic and Statistical Manual* in 2013 in which it called attention to the gradation between troublesome personality patterns and personality disorders along lines reported by Holden (2010).

4. Most researchers now agree that there is a great deal of individual variation in the expression of these patterns and that they are not sharply defined categories. For a discussion of this, see Livesley, et al. (1993); Livesley (2001, 2007); and Skodol, et al. (2005). But others, such as Weston, et al. (2006), argue that these prototypical patterns are still useful ways of thinking about people and talking about them with others. Spitzer, et al. (2008), found that clinicians find it easier to identify personality disorders by using prototypical patterns than by using dimensional traits.

5. Oldham and Morris (1995) have taken the position that "much as high blood pressure represents too much of a good thing, the personality disorders are but extremes of normal human patterns, the stuff of which all our personalities are made." They call adaptive versions of these patterns personality styles and named them as follows: adventurous (antisocial), sensitive (avoidant), mercurial (borderline), conscientious (compulsive), devoted (dependent), dramatic (histrionic), self-confident (narcissistic), vigilant (paranoid), solitary (schizoid), and idiosyncratic (schizotypal).

6. The case for using rankings on Big Five facets to describe these patterns is summed up in a multiauthor book edited

by Costa and Widiger (2002). See also Reynolds and Clark (2001), Lynam and Widiger (2001), Widiger and Samuel (2005), and Widiger and Trull (2007). A table summing up the high and low rankings on Big Five facets for each of these ten patterns appears on page 200 of Widiger and Mullins-Sweatt (2009).

7.   Bobby Fischer's rant is cited by Chun (2002).

8.   Marilyn's recollection is cited by Steinem (1986).

9.   Maccoby (2003) emphasizes the great prevalence of what he calls "productive narcissism" among visionary leaders.

10. Nader (2002).

11. Chamberlain (2004).

12. Dickinson and Pincus (2003); Cain, et al. (2008); and Miller, et al. (2008), emphasize the distinction between grandiose and vulnerable narcissism.

13. Hare (1993) distinguishes psychopaths, whom he identifies with his Psychopathy Checklist, from antisocials as described by DSM-IV. Stout (2005) uses the term sociopath. Millon, et al. (2002), provide a historical review of the terminology and the changing conceptions of this dimensional pattern.

14. Hare (1993).

15. Cresswell and Thomas (2009).

16. Simpson's book became the property of the Goldman family as partial payment for damages they won in a civil suit against him, and they published it in a revised form with the subtitle "confessions of the killer" (Goldman Family, 2007).

17. Hare (1993).

18. Grant, et al. (2004).

19. *Ibid.*

20. Millon (2004).

21. Shawn (2007).

22. Kreisman and Strauss (1989).

23. Grant, et al. (2008).

24. Millon (2004).

25. Oldham and Morris (1995).

26. Beck, et al. (2004), and Beck, et al. (2001). Morf (2006) also emphasizes the importance of understanding a person's view of self and others in making sense of someone's personality.

27. For a more detailed summary of these views of self and others, see Beck, et al. (2004), especially the table on pages 48–49; and Beck, et al. (2001).

Chapter 3

1. Harmon (2006).

2. Ince-Duncan, et al. (2006).

3. Darwin (1859), Chapter 1.

4. Lamason, et al. (2005). However, SLC2A45 is not the only gene that controls skin pigmentation. Sturm (2009) sums up recent information about variants of several other genes that also influence skin pigmentation through a variety of effects on melanin production and display. There is evidence for independent selection of pigmentation genes in East Asian, European, and West African populations, each presumably influenced at least partly by levels of sunlight.

5. Jablonski and Chapin (2000).

6. Galton (1865), cited in Gillham (2001), p. 156.

7. Galton (1875), cited in Gillham (2001), p. 194.

8. Gillham (2001), p. 161.

9. Letter from Charles Darwin to Frances Galton, December 3, 1869, cited in Gillham (2001), p. 169.

10. Plomin, et al. (2008), define heritability as a way of expressing "the extent to which individual differences for a trait in the population can be accounted for by genetic differences among individuals" (p. 82). They give examples of studies that compared Big Five scores of identical and fraternal twins and discuss ways to calculate and interpret heritability based on twin studies and other types of data.

11. Yamagata, et al. (2006), and McCrae and Costa (1997). A study by Riemann, et al. (1997), found even higher heritability. In this study, each twin made a self-assessment with a questionnaire and was also assessed independently by two peers. The results for each twin were then combined to give the cumulative Big Five scores that were used to calculate heritability. The authors concluded that the data derived from three observers provides a better estimate of heritability than data derived solely from one observer, self or peer. Repeated personality testing of twins at an interval of three or more years also increased the accuracy of the results and revealed substantially higher heritability than that based on single measurements (Lyken, 2007). Studies of individual Big Five facets all showed high heritability (Jang, et al., 1996, 1998).

12. Bouchard, et al. (1990).

13. *Ibid.*

14. Getting half of our genes from each parent doesn't translate directly into getting half of each of our parents' personalities. Heritability is a measure of the influence of genes on individual differences in the overall group of people in which it is studied. But it doesn't tell us about the relative genetic and environmental influences on a particular trait, such as high excitement-seeking, in a particular person, such as Jason Dallas. Nevertheless, the studies tell us a lot about the overall influence of genes on our personal tendencies, even if we can't pin down the details in a particular case.

15. Bouchard, et al. (1990), and Bouchard (1994).

16. Harris (1998, 2006) has written two books about the lack of effect of shared family environment on many aspects of personality, and its implications.

17. Plomin and colleagues (Plomin and Daniels [1987], Dunn and Plomin [1991], Plomin, et al. [2001]) have studied the interactions between members of a family that may account for some of the differences in the personalities of siblings. Some of the unique interactions between a parent and a specific child can be attributed to each of their innate tendencies.

18. Turkheimer and Waldron (2000) have emphasized the difficulty of identifying the "nonshared" environmental factors (as opposed to those shared by people who were raised together in the same family) that influence personality. One reason they are hard to identify is that many of them are probably chance events and encounters in the person's life. In Chapter 4, I point out that chance effects on biological processes, such as the migration of neurons in the assembly of the brain and epigenetic

changes in DNA, can also influence personality. These chance effects on biological processes are also included in the category called nonshared environment.

19. New mechanisms of the regulation of gene expression are still being discovered. Some of this regulation is done by specialized proteins called transcription factors; some is done by influences of DNA regions called enhancers, which are not as close to the genes as promoters; and some is done by small specialized bits of another nucleic acid, RNA.

20. Maher (2008); Visscher, et al. (2006); and Yang, et al. (2010). Weedon, et al. (2007), have identified a human gene variant that explains a tiny bit (0.3%) of population variation in height (about two-tenths of an inch).

21. DeFries, et al. (1978).

22. Flint and Mott (2008). Similar studies are being made of the genetic differences in tame and aggressive foxes that were selectively bred for many generations by Dimitry Belayev at the Institute of Cytology and Genetics of the Russian Academy of Sciences (Trut [1999]; Kukekova, et al. [2008]). Studies of brain gene expression in the two lines of foxes have identified many differences (Lindberg, et al. [2005]).

23. These findings should not be taken to mean that the tuning of fearfulness circuits is controlled only by gene variants. Conclusive evidence shows that past experiences also affect fear reactions. Furthermore, such learned changes in the tuning of brain circuits can be as enduring as the tuning controlled by genes. As LeDoux (1998) points out, some of these memories "appear to be indelibly burned into the brain. They are probably with us for life."

24. Turri, et al. (2001); Willis-Owen and Flint (2007); Flint and Mott (2008).

25. Lesch, et al. (1996); Sen, et al. (2006); Canli and Lesch (2007); Caspi, et al. (2010).

26. Hariri, et al. (2002, 2006); Munafo, Brown, et al. (2008). The results are group averages, and there is significant individual variation. See also Oler, et al. (2010).

27. Munafo, Yalcin, et al. (2008) ; Schmidt, et al. (2009).

28. Holden (2008); Ebstein (2006). Many other genes have also been studied in this way.

29. Shifman, et al. (2008); Terraciano, et al. (2010); Calboli, et al. (2010).

30. Lykken, et al. (1992).

31. Several biotechnology companies are competing to develop a machine that will decipher the complete DNA sequence of a person's DNA for as little as $1,000 (Davies [2010]). The companies anticipate that this technology will be used to search for gene variants that influence personality differences.

32. Wolf, et al. (2007); Bell (2007).

33. Buss (1991); MacDonald (1995); Buss and Greiling (1999); Nettle (2005, 2006); Penke, et al. (2007); Ridley (2003); Laland, et al. (2010).

34. Nettle (2005, 2006) has reviewed research on the advantages and disadvantages of high and low rankings on each of the Big Five.

35. Penke, et al. (2007).

36. Maynard Smith (1982) provided a theoretical analysis of the way natural selection tends to maintain a balance between high and low rankings of a trait. His most famous

example, which was based on game theory, concerned the establishment of the relative numbers of aggressive members of a population, which he called Hawks (who always fight over resources), and of nonviolent members, which he called Doves (who never fight). He found that Hawks flourished when they were rare but that as they became more prevalent, they did so much damage to each other that the downside of their aggressiveness began to outweigh its advantages. This resulted in a stable proportion of Hawks and Doves in the population. Once established, this balance tended to persist, and it is referred to as an evolutionarily stable strategy.

Chapter 4

1. Erikson (1963), p. 404.
2. Morris, et al. (2004); Ahmed, et al. (2008). Although fetal testosterone is essential for this process, Wu, et al. (2009), and Junnti, et al. (2010), have shown that the masculinization of neural pathways that control sex-specific behavior actually depends on the enzymatic conversion of fetal testosterone to estrogen by an enzyme in the male mouse brain.
3. Wallace, et al. (2006); Peper, et al. (2007); J. E. Schmitt, et al. (2007); Gilmore, et al. (2010). These differences in the brain structure of identical twins, which may arise partly because of random migration of neurons during brain assembly, may be responsible for some of the personality differences of the members of a twin pair.
4. Hensch (2004).
5. Doupe and Kuhl (1999).

6. Lenneberg (1967); Doupe and Kuhl (1999); Perani and Abutalebi (2005). In the case of human language, the window is always kept a little open, so new languages can still be learned in adulthood, but with progressively greater difficulty. Because this window never closes completely, some prefer to call this a "sensitive period" rather than a critical period.

7. Harris (1998) emphasizes the well-known fact that the young children of immigrants whose parents speak English with a foreign accent learn to speak like their peers instead of their parents. She takes this as strong evidence that the social environment that children care about, and are mainly molded by, is the environment provided by their peers rather than their parents.

8. Thomas, et al. (1963); Chess and Thomas (1986).

9. Chess and Thomas (1986).

10. Kagan (1994), p. 135.

11. Schwartz, et al. (2003).

12. The general conclusion that childhood behavior is somewhat predictive of behavior in later life is supported by the longitudinal studies of many investigators and reported in publications by Block (1993); Block and Block (2006); Caspi (2000); Caspi, et al. (2003); Dennissen, et al. (2008); Hampson and Goldberg (2006); Mischel, et al. (1988); Shiner (2000, 2005); Shiner, et al. (2002, 2003).

13. Goldstein, et al. (2006).

14. DiLalla and Gottesman (1989); Taylor, et al. (2000). Why do children vary greatly in their adherence to their earlier behavioral paths? Kagan (1994) believes that parenting makes a big difference. But Harris (1998) has challenged this belief in the importance of parenting. She

points, instead, to the powerful influence of peers. And she goes further. Instead of simply shrugging off opinions such as Kagan's as unverified but harmless, Harris believes that it "has put a terrible burden of guilt on parents unfortunate enough to have ... for some reason failed to produce a happy, smart, well-adjusted, self-confident person. Not only must these parents suffer the pain of having a child who is difficult to live with or who fails in some other way to live up to the community's standards: they must also bear the community's opprobrium." (Harris, 1998, p. 352)

15. Kendler; Gardner, et al. (2008).

16. Moffitt (2005); Mealey (1995).

17. Moffitt (2005).

18. Miles and Carey (1997); Rhee and Waldman (2002).

19. Miles and Carey (1997); Rhee and Waldman (2002); Moffitt (2005).

20. Caspi, et al. (2002).

21. Edwards, et al. (2003).

22. Caspi, et al. (2002).

23. Plomin, et al. (2001). Evidence (Kendler, Jacobson, et al. [2007, 2008]) also indicates that a child's innate tendencies influence his or her selection of peers and that this, too, may contribute to the development of an antisocial pattern.

24. Meyer-Lindenberg, et al. (2006); Buckholtz and Meyer-Lindenberg (2008); Buckholtz, et al. (2008).

25. Sabol, et al. (1998).

26. Caspi, et al. (2002).

27. *Ibid.*

28. Foley, et al. (2004) ; Kim-Cohen, et al. (2006).

29. Ducci, et al. (2008).

30. Meyer-Lindenberg, et al. (2006); Buckholtz and Meyer-Lindenberg (2008); Buckholtz, et al. (2008).

31. Meaney (2001).

32. Weaver, et al. (2004); Meaney and Szyf (2005); Buchen (2010).

33. Zhang and Meaney (2010). Methylation and demethylation of DNA are not the only epigenetic changes that influence gene expression. Epigenetic changes also occur by chemical modifications of histones that are associated with DNA in chromosomes (Kouzarides, 2007), and changes in acetylation of histones in specific brain cells have been shown to control behavioral adaptations to emotional stimuli (Renthal, et al. [2007]).

34. Kaffman and Meaney (2007); McGowan, et al. (2009); Heim and Nemeroff (2001); Rinne, et al. (2002). Tottenham and Sheridan (2010) review some effects of early adverse social environments on behavior later in life.

35. Fraga, et al. (2005); Haque, et al. (2009); Kaminsky, et al. (2009).

36. Feinberg and Irizzary (2010) have proposed that some of these epigenetic differences arise stochastically (randomly) instead of in response to specific environmental influences, and that these random variations provide variability that may increase fitness in particular environments. This "stochastic epigenetic variation" not only may explain some of the methylation differences observed in the DNA of identical twins. It also may contribute to that ill-defined entity called a "nonshared environment" (Plomin, et al. [2008]; Turkheimer and Waldron [2000]) that has been put forth as the explanation for their personality differences.

37. Morris, et al. (2004); Sisk and Foster (2004); Romeo (2003).

38. Arnold, et al. (2003); Ahmed, et al. (2008); Sisk and Zehr (2005).

39. Blakemore (2008); Steinberg (2010).

40. Arnold, et al. (2003); Morris, et al. (2008).

41. Plomin, et al. (1997); Petrill, et al. (2004). Shaw, et al. (2006), describe the relationship between intellectual ability and changes in cortical thickness during adolescence.

42. Haworth, et al. (2009).

43. Giedd, et al. (1999); Sowell (2003); Thompson, et al. (2005); Shaw, et al. (2008); Giedd (2008); Ernst and Mueller (2008).

44. Fair, et al. (2008); Ernst and Mueller (2008); Dosenbach, et al. (2010).

45. Harris (1998, 2006).

46. Kendler, et al. (2007); Kendler, Jacobson, et al. (2008).

47. Sowell, et al. (2003); Thompson, et al. (2005).

48. Bartzokis, et al. (2001).

49. Dosenbach, et al. (2010).

50. Roberts and DelVecchio (2000) found continuing stabilization of a person's Big Five rankings until about the age of 50, whereas McCrae and Costa (2003) argue that there really isn't much change in a person's relative rankings after the age of 30. Studies of average scores of groups of people at different ages (as opposed to rank order of individuals) show an increase in Conscientiousness and Agreeableness in the overall population into old age

(Srivastava, et al. [2003]; Roberts, et al. [2006]; Costa and McCrae [2006]), as well as some other changes.

51. Roberts and Caspi (2003) present persuasive arguments for what they call "the cumulative continuity model of personality development" that emphasizes the contribution of sustained person–environment transactions to the stability of adult personality. McCrae and Costa (1994) point out the great value of the stabilization of personality in young adulthood. As they put it, "Because personality is stable, life is to some extent predictable. People can make vocational and retirement choices with some confidence that their current interests and enthusiasms will not desert them. They can choose mates and friends with whom they are likely to remain compatible.... They can learn which coworkers they can depend on, and which they cannot. The personal and social utility of social stability is enormous."

Chapter 5

1. Franklin's autobiography is available without charge at several online sites, including www.earlyamerica.com/lives/franklin/. All the quotes attributed to him are from his autobiography, unless otherwise indicated.

2. For facts and interpretations of Franklin's life, I have relied mainly on Isaacson (2003).

3. Allport (1961), p. 31.

4. David Hume, the Scottish philosopher, was among the first to emphasize the link between morals and emotions. As he put it in 1739, "Morals excite passions, and produce or prevent actions. Reason itself is utterly impotent in

this particular. The rules of morality, therefore, are not conclusions of our reason."

5. Darwin (1871), Chapter 4.

6. Trivers (1971).

7. de Waal (1996); Wright (1994); Pinker (2002); Ridley (1996).

8. de Waal (2008).

9. A specialized group of neurons, called mirror neurons because they are activated when we mirror the behavior of others, may play an important part in the brain circuits that participate in empathy (Gallese [2001]; Decety and Jackson [2004]; Preston and de Waal [2002]).

10. Haidt (2003) and Tangney, et al. (2007), have reviewed research on moral emotions.

11. Haidt (2003).

12. Trivers (1971). Also see Boyd, et al. (2010), on the importance of punishment and negative emotions in sustaining cooperation.

13. Darwin (1871), Chapter 4. Laland, et al. (2010), have argued that culture has also shaped the human genome.

14. Peterson and Seligman (2004).

15. Cloninger, et al. (1993); Cloninger (2004).

16. Shweder, et al. (1997).

17. Shweder (1994) gives many examples: "On the basis of the historical and ethnographic record we know that different people in different times and places have found it quite natural to be spontaneously appalled, outraged, indignant, proud, disgusted, guilty, and ashamed by all sorts of things: masturbation, homosexuality, sexual abstinence, polygamy, abortion, circumcision, corporal punishment, capital

punishment, Islam, Christianity, Judaism, capitalism, democracy, flag burning, miniskirts, long hair, no hair, alcohol consumption, meat eating, medical inoculations, atheism, idol worship, divorce, widow remarriage, arranged marriage, romantic love marriage, parents and children sleeping in the same bed, parents and children not sleeping in the same bed, women being allowed to work, women not being allowed to work."

18. Covey (1989).

19. Charles Angoff, cited by Isaacson (2003), p. 483.

20. Isaacson (2003), p. 476.

21. Franklin, cited by Isaacson, p. 87.

22. Isaacson (2003), p. 487.

23. Ellis (2003).

Chapter 6

1. Bruner (1985, 1990).

2. McAdams (1993), p. 266.

3. Erikson (1980).

4. McCrae and Costa (2003, p. 191) call these characteristic ways of dealing with the world characteristic adaptations: "They are *characteristic* because they reflect the operation of enduring personality traits, and they are *adaptations* because they are shaped in response to the demands and opportunities offered by the environment." For other views of the role of characteristic adaptations in identity and personality, see McAdams and Pals (2006), McAdams and Olson (2009), Roberts and Robins (2000), and Bleidorn, et al. (2010).

5. Friedman (1990).

6. Kelly (2010) is my main source for the details of Oprah's life.

7. Kelly (2010), p. 34.

8. Cited by Kelly (2010), p. 40.

9. Cited by Kelly (2010), p. 3.

10. McAdams (1993).

11. Bandura (1982).

12. Erikson (1958), p. 111.

13. Kelly (2010), p. 24.

14. Kelly (2010), p. 34.

15. Isaacson (2003), p. 2.

16. Writers of memoirs also recognize the importance of inventiveness. In *Inventing the Truth*, Zinsser (1998, p. 6) tells us that "memoir writers must manufacture a text, imposing narrative order on a jumble of half-remembered events. With that feat of manipulation they arrive at a truth that is theirs alone."

17. Erikson (1980).

18. To get the full flavor, you can watch the video of Jobs's commencement address on You Tube: www.youtube.com/watch?v=UF8uR6Z6KLc.

19. Elkind (2008). All the quotes that follow are by Elkind or are cited by him.

20. McAdams (1993).

Chapter 7

1. Kluckhohn and Murray (1953).

2. Gordon Allport was well aware of our reliance on this
    natural intuitive ability as the starting point for a conscious
    assessment. To him, learning to analyze a personality was
    like learning to analyze a piece of music. As he explained
    it, "No one learns to hear the tonal pattern of a symphony,
    but we can be taught to listen to it and to look for
    significant features. Most instruction in life is devoted to
    analysis, to giving knowledge about, and to building up
    a store of available inferences. We cannot teach another
    to perceive the unity of an object (it is simply there),
    but we can teach so that his associational equipment is
    enriched. And so it is with personality: We cannot teach
    understanding of pattern, but we can call attention to
    detail, as well as to laws, principles, generalizations which
    can sharpen comprehension through comparison and
    inference." (Allport [1961], p. 547)

3. A good example of a nuanced description of a Big Five
    tendency comes from Lev Grossman's (2010) Person of the
    Year article about Facebook founder Mark Zuckerberg, in
    which he considers Zuckerberg's ranking on Extraversion.
    Using the Big Five framework helps me appreciate,
    integrate, and remember Grossman's insightful assessment:

    "Zuckerberg has often—possibly always—been described
    as remote and socially awkward, but that's not quite right.
    True: holding a conversation with him can be challenging.
    He approaches conversation as a way of exchanging data
    as rapidly and efficiently as possible, rather than as a
    recreational activity undertaken for its own sake. He is
    formidably quick and talks rapidly and precisely, and if he

has no data to transmit, he abruptly falls silent. ('I usually don't like things that are too much about me' was how he began our first interview.) He cannot be relied on to throw the ball back or give you encouraging facial cues. His default expression is a direct and slightly wide-eyed stare that makes you wonder if you've got a spider on your forehead....

"In spite of all that—and this is what generally gets left out—Zuckerberg is a warm presence, not a cold one. He has a quick smile and doesn't shy away from eye contact. He thinks fast and talks fast, but he wants you to keep up. He exudes not anger or social anxiety, but a weird calm. When you talk to his coworkers, they're so adamant in their avowals of affection for him and in their insistence that you not misconstrue his oddness that you get the impression it's not just because they want to keep their jobs. People really like him....

"The reality is that Zuckerberg isn't alienated, and he isn't a loner. He's the opposite. He's spent his whole life in tight, supportive, intensely connected social environments: first in the bosom of the Zuckerberg family, then in the dorms at Harvard, and now at Facebook, where his best friends are his staff, there are no offices, and work is awesome. Zuckerberg loves being around people. He didn't build Facebook so he could have a social life like the rest of us. He built it because he wanted the rest of us to have his."

4. Funder and Sneed (1993).
5. Clinton (2004).
6. Obama (1995).
7. *Ibid.*
8. Mundy (2007).

9.   Obama (2006).

10. Roberts and DelVecchio (2000); McCrae and Costa (2003).

11. Roberts and Caspi (2003).

12. Avdi and Gorgaca (2007); Adler, et al. (2008); Salvatore, et al. (2004); Wilson (2002).

References

Adler, J. M., L. M. Skalina, and D. P. McAdams. "The Narrative Reconstruction of Psychotherapy and Psychological Health." *Psychotherapy Research* 18 (2008): 719–734.

Ahmed, E. I., J. L. Zehr, K. L. Schulz, K. M. Schulz, B. M. Lorenz, L. L. DonCarlos, and C. L. Sisk. "Pubertal Hormones Modulate the Addition of New Cells to Sexually Dimorphic Brain Regions." *Nature Neuroscience* 11 (2008): 995–997.

Allport, G. W. *Pattern and Growth in Personality*. New York: Holt, Rinehart and Winston, 1961.

Allport, G. W., and H. S. Odbert. "Trait Names: A Psycholexical Study." *Psychological Monographs* 47 (1936): Whole No. 211.

American Psychiatric Association. *Diagnostic and Statistical Manual of Mental Disorders*, 4th ed. (DSM-IV). Washington, D.C.: American Psychiatric Association Press, 1994.

Arnold, A. P., E. F. Rissman, and G. J. DeVries. "Two Perspectives on the Origin of Sex Differences in the Brain." *Annals of New York Academy of Science* 1,007 (2003): 176–188.

Avdi, E., and E. Georgaca. "Narrative Research in Psychotherapy: A Critical Review." *Psychology and Psychotherapy: Theory, Research and Practice* 80 (2007): 407–419.

Bandura, A. "The Psychology of Chance Encounters and Life Paths." *American Psychologist* 37 (1982): 747–755.

Bargh, J. A., and E. L. Chartrand. "The Unbearable Automaticity of Being." *American Psychologist* 54 (1999): 462–479.

Bargh, J. A., and E. L. Williams. "The Automaticity of Social Life." *Current Directions in Psychological Science* 15 (2006): 1–5.

Bartzokis, G., M. Beckson, P. H. Lu, K. H. Neuchterlein, N. Edwards, and J. Mintz. "Age-Related Changes in Frontal and Temporal Lobe Volumes in Men." *Archives of General Psychiatry* 58 (2001): 461–465.

Beck, A. T., A. C. Butler, G. K. Brown, K. K. Dahlsgaard, C. F. Newman, and J. S. Beck. "Dysfunctional Beliefs Discriminate Personality Disorders." *Behaviour Research and Therapy* 39 (2001): 1,213–1,225.

Beck, A. T., A. Freeman, D. D. David, and Associates. *Cognitive Therapy of Personality Disorders.* New York: Guilford Press, 2004.

Bell, A. M. "Animal Personalities." *Nature* 447 (2007): 539–540.

Blakemore, S. J. "The Social Brain in Adolescence." *Nature Reviews Neuroscience* 9 (2008): 267–277.

Bleidorn, W., C. Kandler, U. T. Hulsheger, R. Riemann, A. Angleitner, and F. M. Spinath. "Nature and Nurture of the Interplay Between Personality Traits and Major Life Goals." *Journal of Personality and Social Psychology* 99 (2010): 366–379.

Block, J. "Studying Personality the Long Way." In *Studying Lives Through Time: Personality and Development.* Edited by D. C. Funder, R. D. Parke, and C. Tomlinson-Keasy. Washington, D.C.: American Psychological Association, 1993.

Block, J., and J. H. Block. "Venturing a 30-Year Longitudinal Study." *American Psychologist* 61 (2006): 315–327.

Bouchard, T. J. Jr. "Genes, Environment, and Personality." *Science* 264 (1994): 1,700–1,701.

Bouchard, T. J. Jr., D. T. Lykken, M. McGue, N. L. Segal, and A. Tellegen. "Sources of Human Psychological Differences: The Minnesota Study of Twins Reared Apart." *Science* 250 (1990): 223–228.

Boyd, R., H. Gintis, and S. Bowles. "Coordinated Punishment of Defectors Sustains Cooperation and Can Proliferate When Rare." *Science* 328 (2010): 617–620.

Bruner, J. *Actual Minds, Possible Worlds.* Cambridge, MA: Harvard University Press, 1985.

Bruner, J. *Acts of Meaning: Four Lectures on Mind and Culture.* Cambridge, MA: Harvard University Press, 1990.

Buchanan, T., J. A. Johnson, and L. R. Goldberg. "Implementing a Five-Factor Personality Inventory for Use on the Internet." *European Journal of Psychological Assessment* 21 (2005): 115–127.

Buchen, L. "Neuroscience: In Their Nurture." *Nature* 467 (2010): 146–148.

Buckholtz, J. W., and A. Meyer-Lindenberg. "MAOA and the Neurogenetic Architecture of Human Aggression." *Trends in Neurosciences* 31 (2008): 120–129.

Buckholtz, J. W., J. H. Callicot, B. Kolachana, A. R. Hariri, T. E. Goldberg, M. Genderson, M. F. Egan, V. S. Mattay, D. R. Weinberger, and A. Meyer-Lindenberg.

"Genetic Variation in MAOA Modulates Ventromedial Prefrontal Circuitry Mediating Individual Differences in Human Personality." *Molecular Psychiatry* 13 (2008): 313–324.

Buss, D. M. "Evolutionary Personality Psychology." *Annual Review of Psychology* 42 (1991): 459–491.

Buss, D. M., and H. Greiling. "Adaptive Individual Differences." *Journal of Personality* 67 (1999): 209–243.

Cain, N. M., A. L. Pincus, and E. B. Ansell. "Narcissism at the Crossroads: Phenotypic Description of Pathological Narcissism Across Clinical Theory, Social/Personality Psychology and Psychiatric Diagnosis." *Clinical Psychology Review* 28 (2008): 638–656.

Calboli, F. C. F., F. Tozzi, N. W. Galwey, A. Antoniades, V. Moosr, M. Preisig, P. Vollenweider, D. Waterworth, G. Waeber, M. R. Johnson, P. Muglia, and D. J. Balding. "A Genome-wide Association Study of Neuroticism in a Population-Based Sample." *PLoS One* 5 (2010): e11,504.

Canli, T., and K. P. Lesch. "Long Story Short: The Serotonin Transporter in Emotion Regulation and Social Cognition." *Nature Neuroscience* 10 (2007): 1,103–1,109.

Caspi, A. "The Child Is Father of the Man: Personality Correlates from Childhood to Adulthood." *Journal of Personality and Social Psychology* 78 (2000): 158–172.

Caspi, A., A. R. Hariri, A. Holmes, R. Uher, and T. E. Moffitt. "Genetic Sensitivity to the Environment: The Case of the Serotonin Transporter Gene and Its Implications for Studying Complex Diseases and Traits." *American Journal of Psychiatry* 167 (2010): 509–527.

Caspi, A., H. Harrington, B. Milne, J. W. Amell, R. F. Theodore, and T. E. Moffitt. "Children's Behavioral Styles at Age 3 Are Linked to Adult Personality Traits at Age 26." *Journal of Personality* 71 (2003): 495–513.

Caspi, A., J. McClay, T. E. Moffitt, J. Mill, J. Martin, I. W. Craig, A. Taylor, and R. Poulton. "Role of Genotype in the Cycle of Violence in Maltreated Children." *Science* 297 (2002): 851–854.

Caspi, A., and T. E. Moffitt. "Gene–Environment Interactions in Psychiatry: Joining Forces with Neuroscience." *Nature Reviews Neuroscience* 7 (2006): 583–590.

Caspi, A., and B. Roberts. "Personality Continuity and Change Across the Life Course." In *Handbook of Personality: Theory and Research*. Edited by L. A. Pervin. New York: Guilford Press, 1999.

Caspi, A., B. W. Roberts, and R. L. Shiner. "Personality Development: Stability and Change." *Annual Review of Psychology* 56 (2005): 453–484.

Chamberlain, L. "The Dark Side of Ralph Nader." Salon.com. July 1, 2004.

Chess, S., and A. Thomas. *Temperament in Clinical Practice*. New York: Guilford Press, 1986.

Chun, R. "Bobby Fischer's Pathetic Endgame." *The Atlantic,* December 2002.

Clinton, B. *My Life*. New York: Alfred A. Knopf, 2004.

Cloninger, C. R. *Feeling Good: The Science of Well-Being*. New York: Oxford University Press, 2004.

Cloninger, C.R., D. M. Svrakic, and T. R. Przybeck. "A Psychobiological Model of Temperament and Character." *Archives of General Psychiatry* 50 (1993): 975–990.

Costa, P. T. Jr., and R. R. McCrae. *NEO PI-R: Professional Manual*. Odessa, FL: Psychological Assessment Resources, 1992.

Costa, P. T. Jr., and R. R. McCrae. "Age Changes in Personality and Their Origins: Comment on Roberts, Walton, and Viechtbauer." *Psychological Bulletin* 132 (2006): 26–28.

Costa, P. T. Jr., A. Terracciano, and R. R. McCrae. "Gender Differences in Personality Traits Across Cultures: Robust and Surprising Findings." *Journal of Personality and Social Psychology* 81 (2001): 322–331.

Costa, P. T. Jr., and T. A. Widiger, eds. *Personality Disorders and the Five-Factor Model of Personality,* 2nd ed. Washington, D.C.: American Psychological Association, 2002.

Covey, S. R. *The 7 Habits of Highly Effective People*. New York: Simon & Schuster, 1989.

Craik, K. H., R. T. Hogan, and R. N. Wolfe, eds. *Fifty Years of Personality Psychology*. New York: Plenum Press, 1993.

Creswell, J., and L. Thomas. "The Talented Mr. Madoff." *The New York Times.* January 25, 2009.

Darwin, C. *On the Origin of Species.* 1859. www.literature.org/authors/darwin-charles/the-origin-of-species/.

Darwin, C. *The Descent of Man.* 1871. www.gutenberg.org/ebooks/2300.

Davies, K. *The $1,000 Genome: The Revolution in DNA Sequencing and the New Era of Personalized Medicine*. New York: Free Press, 2010.

Decety, J., and P. L. Jackson. "The Functional Architecture of Human Empathy," *Behavioral and Cognitive Neuroscience Reviews* 3 (2004): 71–100.

DeFries, J., M. Gervais, and E. Thomas. "Response to 30 Generations of Selection for Open-Field Activity in Laboratory Mice." *Behavior Genetics* 8 (1978): 3–13.

Dennissen, J. J. A., J. B. Aspendorpf, and M. A. G. van Aken. "Childhood Personality Predicts Long-Term Trajectories of Shyness and Aggressiveness in the Context of Demographic Transitions in Emerging Adulthood." *Journal of Personality* 76 (2008): 67–99.

Denissen, J. J. A., and L. Penke. "Motivational Individual Reaction Norms Underlying the Five-Factor Model of Personality: First Steps Towards a Theory-Based Conceptual Framework." *Journal of Research in Personality* 42 (2008): 1,285–1,302.

de Waal, F. B. M. *Good Natured: The Origins of Right and Wrong in Humans.* Cambridge, MA: Harvard University Press, 1996.

de Waal, F. B. M. "Putting the Altruism Back into Altruism: The Evolution of Empathy." *Annual Review of Psychology* 59 (2008): 279–300.

Dickinson, K. A., and A. L. Pincus. "Interpersonal Analysis of Grandiose and Vulnerable Narcissism." *Journal of Personality Disorders* 17 (2003): 188–207.

Digman, J. M. "Personality Structure: Emergence of the Five-Factor Model." *Annual Review of Psychology* 41 (1990): 417–440.

DiLalla, L. F., and I. Gottesman. "Heterogeneity of Causes for Delinquency and Criminality: Lifespan Perspectives." *Developmental Psychopathology* 1 (1989): 339–349.

Dosenbach, N. U., B. Nardos, A. L. Cohen, D. A. Fair, J. D. Power, J. A. Church, S. M. Nelson, G. S. Wig, A. C. Vogel, C. N. Lessov-Schlagger, K. A. Barnes, J. W. Dubis, E. Feczko, R. S. Coalson, J. R. Pruett, D. M. Barch, S. E. Petersen, and B. L. Schlaggar. "Prediction of Individual Brain Maturity Using fMRI." *Science* 329 (2010): 1,358–1,361.

Doupe, A. J., and P. K. Kuhl. "Birdsong and Human Speech: Common Themes and Mechanisms." *Annual Review Neuroscience* 22 (1999): 567–631.

Dowd, M. "Captain Obvious Learns the Limits of Cool." *The New York Times.* January 9, 2010.

Ducci, F., M. A. Enoch, C. Hodgkinson, K. Xu, M. Catena, R. W. Robin, and D. Goldman. "Interaction Between a Functional MAOA Locus and Childhood Sexual Abuse Predicts Alcoholism and Antisocial Personality Disorder in Adult Women." *Molecular Psychiatry* 13 (2008): 334–347.

Dunn, J., and R. Plomin. "Why Are Siblings So Different? The Significance of Differences in Sibling Experiences Within the Family." *Family Process* 30 (1991): 271–283.

Eaves, L. J., E. C. Prom, and J. L. Silberg. "The Mediating Effect of Parental Neglect on Adolescent and Young Adult Anti-sociality: A Longitudinal Study of Twins and Their Parents." *Behavioral Genetics* 40 (2010): 425–437.

Ebstein, R. P. "The Molecular Genetic Architecture of Human Personality: Beyond Self-Report Questionnaires." *Molecular Psychiatry* 11 (2006): 427–445.

Edwards, V. J., G. W. Holden, V. J. Felitti, and R. F. Anda. "Relationship Between Multiple Forms of Childhood Maltreatment and Adult Mental Health in Community Respondents: From the Adverse Childhood Experiences Study." *American Journal of Psychiatry* 160 (2003): 1,453–1,460.

Elkind, P. "The Trouble with Steve Jobs." *CNN Money.* March 5, 2008.

Ellis, J. "Benjamin Franklin: The Many-Minded Man." *The New York Times.* July 6, 2003.

Epstein, S., R. Pacini, V. Denes-Raj, and H. Heier. "Individual Differences in Intuitive–Experiential and Analytical–Rational Thinking Styles." *Journal of Personality and Social Psychology* 71 (1996): 390–405.

Erikson, E. H. *Young Man Luther.* New York: Norton, 1958.

Erikson, E. H. *Childhood and Society,* 2nd ed. New York: Norton, 1963.

Erikson, E. H. *Identity and the Life Cycle.* New York: Norton, 1980.

Ernst, M., and S. C. Mueller. "The Adolescent Brain: Insights from Functional Neuroimaging Research." *Developmental Neurobiology* 68 (2008): 729–743.

Eysenck, H. J. *The Structure of Human Personality.* London: Methuen, 1965.

Fair, D. A., A. L. Cohen, N. U. Dosenbach, J. A. Church, F. M. Miezin, D. M. Barch, M. E. Raichle, S. E. Petersen, and B. L. Schlaggar. "The Maturing Architecture of the Brain's Default Network." *Proceedings of the National Academy of Sciences* 105 (2008): 4,028–4,032.

Feinberg, A. P., and R. A. Irizarry. "Stochastic Epigenetic Variation as a Driving Force of Development, Evolutionary Adaptation, and Disease." *Proceedings of the National Academy of Sciences* 107 (2010): 1,757–1,764.

Fiske, S. T. A., J. C. Cuddy, and P. Glick. "Universal Dimensions of Social Cognition: Warmth and Competence." *Trends in Cognitive Sciences* 11 (2006): 77–83.

Flint, J., and R. Mott. "Applying Mouse Complex-Trait Resources to Behavioural Genetics." *Nature* 456 (2008): 724–727.

Foley, D. L., L. J. Eaves, B. Wormley, J. L. Silberg, H. H. Maes, J. Kuhn, and B. Riley. "Childhood Adversity, MAOA Genotype, and Risk for Conduct Disorder." *Archives of General Psychiatry* 61 (2004): 738–744.

Fraga, M. F., E. Ballestar, M. F. Paz, S. Ropero, F. Setien, M. L. Ballestar, D. Heine-Suner, J. C. Cigudosa, M. Urioste, J. Benitez, M. Boix-Chronet, A. Sanchez-Aguilra, C. Ling, E. Carlsson, P. Poulsen, A. Vaag, Z. Stephan, T. D. Spector, Y. Z. Wu, C. Plass, and M. Esteller. "Epigenetic Differences Arise During the Lifetime of Monozygotic Twins." *Proceedings of the National Academy of Sciences* 102 (2005): 10,604–10,609.

Friedman, L. J. *Identity's Architect: A Biography of Erik H. Erikson.* New York: Scribner, 1990.

Frith, C. D., and U. Frith. "Implicit and Explicit Processes in Social Cognition." *Neuron* 60 (2008): 503–510.

Funder, D. C. "On the Accuracy of Personality Judgment: A Realistic Approach." *Psychological Review* 102 (1995): 652–670.

Funder, D. C. "Towards a Resolution of the Personality Triad: Persons, Situations, Behaviors." *Journal of Research in Personality* 40 (2006): 21–34.

Funder, D. C., and C. D. Sneed. "Behavioral Manifestations of Personality: An Ecological Approach to Judgmental Accuracy." *Journal of Personality and Social Psychology* 64 (1993): 479–490.

Gallese, V. "The 'Shared Manifold' Hypothesis: From Mirror Neurons to Empathy." *Journal of Consciousness Studies* 8 (2001): 33–50.

Giedd, J. N. "The Teen Brain: Insights from Neuroimaging." *Journal of Adolescent Health* 42 (2008): 335–343.

Giedd, J. N., J. Blumenthal, N. O. Jeffries, F. X. Castellanos, H. Liu, A. Zijdenbos, T. Paus, A. C. Evans, and J. L. Rapoport. "Brain Development During Childhood and Adolescence: A Longitudinal MRI Study." *Nature Neuroscience* 2 (1999): 861–863.

Gilmore, J. H., J. E. Schmitt, R. C. Knickmeyer, J. K. Smith, W. Lin, M. Styner, G. Gerig, and M. C. Neale. "Genetic and Environmental Contributions to Neonatal Brain Structure: A Twin Study." *Human Brain Mapping* 31 (2010): 1,174–1,182.

Gigerenzer, G. *Gut Feelings: The Intelligence of the Unconscious.* New York: Viking, 2007.

Gigerenzer, G., and D. G. Goldstein. "Reasoning the Fast and Frugal Way: Models of Bounded Rationality." *Psychological Review* 103 (1996): 650–669.

Gillham, N. W. *A Life of Sir Francis Galton: From African Exploration to the Birth of Eugenics.* New York: Oxford University Press, 2001.

Gladwell, M. *Blink: The Power of Thinking Without Thinking.* New York: Little Brown and Company, 2005.

Goldberg, L. R. "An Alternative 'Description of Personality': The Big-Five Factor Structure." *Journal of Personality and Social Psychology* 59 (1990): 1,216–1,229.

Goldberg, L. R. "The Development of Markers for the Big Five Factor Structure." *Psychological Assessment* 4 (1992): 26–42.

Goldberg, L. R. "The Structure of Phenotypic Personality Traits." *American Psychologist* 48 (1993): 26–34.

Goldberg, L. R., J. A. Johnson, H. W. Eber, R. Hogan, M. C. Ashton, C. R. Cloninger, and H. G. Gough. "The International Personality Item Pool and the Future of Public Domain Personality Measures." *Journal of Research in Personality* 40 (2006): 84–96.

Goldman Family. *If I Did It: Confessions of the Killer.* New York: Beaufort Books, 2007.

Goldstein, R. B., B. F. Grant, W. J. Ruan, S. M. Smith, and T. D. Saha. "Antisocial Personality Disorder with Childhood vs. Adolescent-Onset Conduct Disorder." *The Journal of Nervous and Mental Disease* 194 (2006): 667–675.

Gosling, S. D., S. Vazire, S. Srivastava, and O. P. John. "Should We Trust Web Based Studies?: A Comparative Analysis of Six Preconceptions About Internet Questionnaires." *American Psychologist* 59 (2004): 93–104.

Grant, B. F., D. S. Hasin, F. S. Stinson, D. A. Dawson, S. P. Chou, W. J. Ruan, and R. P. Pickering. "Prevalence, Correlates, and Disability of Personality Disorders in the United States: Results from the National Epidemiologic Survey on Alcohol and Related Conditions." *Journal of Clinical Psychiatry* 65 (2004): 948–958.

Grant, B. F., S. P. Chou, R. B. Goldstein, B. Huang, F. S. Stinson, T. D. Saha, S. M. Smith, D. A. Dawson, A. J. Pulay, R. P. Pickering, and W. J. Ruan. "Prevalence, Correlates, Disability, and Comorbidity of DSM-IV Borderline Personality Disorder." *Journal of Clinical Psychiatry* 69 (2008): 533–545.

Grossman, L. "Mark Zuckerberg." *Time.* December 15, 2010.

Haidt, J. "The Moral Emotions." In *Handbook of Affective Sciences.* Edited by R. J. Davidson, K. R. Scherer, and H. H. Goldsmith. New York: Oxford University Press, 2003.

Hampson, S. E., and L. R. Goldberg. "A First Large Cohort Study of Personality Trait Stability over the Years Between Elementary School and Midlife." *Journal of Personality and Social Psychology* 91 (2006): 763–779.

Haque, F. N., I. I. Gottesman, and A. H. C. Wong. "Not Really Identical: Epigenetic Differences in Monozygotic Twins and Implications for Twin Studies in Psychiatry." *American Journal of Medical Genetics*. Part C, 151C (2009): 136–141.

Hare, R. D. *Without Conscience: The Disturbing World of the Psychopaths Among Us*. New York: Guilford Press, 1993.

Hariri, A. R., V. S. Mattay, A. Tessitore, B. Kolachan, F. Fera, D. Goldman, M. F. Egan, and D. R. Weinberger. "Serotonin Transporter Genetic Variation and the Response of the Human Amygdala." *Science* 297 (2002): 400–403.

Hariri, A. R., E. M. Drabant, and D. R. Weinberger. "Imaging Genetics: Perspectives from Studies of Genetically Driven Variation in Serotonin Function and Corticolimbic Affective Processing." *Biological Psychiatry* 59 (2006): 888–897.

Harmon, A. "That Wild Streak? Maybe It Runs in the Family?" *The New York Times*. June 15, 2006.

Harris, J. R. *The Nurture Assumption: Why Children Turn Out the Way They Do*. New York: Free Press, 1998.

Harris, J. R. *No Two Alike: Human Nature and Human Individuality*. New York: Norton, 2006.

Haworth, C. M. A., M. J. Wright, M. Luciano, N. G. Martin, E. J. C. de Geus, C. E. M. van Eijsterveldt, M. Bartles, D. Posthuma, D. I. Boomsma, O. S. P. Davis, Y. Kovas, R. P. Corley, J. C. DeFries, J. K. Hewitt, R. K. Olson, S. A. Rhea, S. J. Wadsworth, W. G. Iacono, M. McGue, L. A. Thompson, S. A. Hart, S. A. Petrill, D. Lubinski, and R. Plomin. "The Heritability of General Cognitive Ability Increases Linearly from Childhood to Young Adulthood." *Molecular Psychiatry* advance online publication. June 2, 2009.

Hegelson, V., and H. L. Fritz. "The Implications of Unmitigated Agency and Unmitigated Communion for Domains of Problem Behavior." *Journal of Personality* 68 (2000): 1,031–1,057.

Heim, C. and C. B. Nemeroff. "The Role of Childhood Trauma in the Neurobiology of Mood and Anxiety Disorders: Preclinical and Clinical Studies." *Biological Psychiatry* 49 (2001): 1,023–1,039.

Hensch, T. K. "Critical Period Regulation." *Annual Review of Neuroscience* 27 (2004): 549–579.

Hitchens, C. *No One Left to Lie To: The Triangulations of William Jefferson Clinton*. New York: Verso, 1999.

Holden, C. "Parsing the Genetics of Behavior." *Science* 322 (2008): 892–895.

Holden, C. "APA Seeks to Overhaul Personality Disorder Diagnosis." *Science* 327 (2010): 1,314.

Hume, D. *A Treatise of Human Nature: Being an Attempt to Introduce the Experimental Method of Reasoning into Moral Subjects*. Mineola, NY: Dover Philosophical Classics, 1739.

Ince-Dunn, G., B. Hall, S. C. Hu, B. Ripley, R. L. Huganir, J. M. Olsosn, S. J. Tapscott, and A. Ghosh. "Regulation of Thalamocortical Patterning and Synaptic Maturation by NeuroD2." *Neuron* 49 (2006): 639–651.

Isaacson, W. *Benjamin Franklin: An American Life*. New York: Simon & Schuster, 2003.

Jablonski, N. G., and G. Chaplin. "The Evolution of Human Skin Coloration." *Journal of Human Evolution* 39 (2000): 57–106.

Jaffee, S. R., A. Caspi, T. E. Moffitt, and A. Taylor. "Victim of Abuse to Antisocial Child: Evidence of an Environmentally Mediated Process." *Journal of Abnormal Psychology* 113 (2004): 44–55.

Jang, K. L., W. J. Livesley, and P. A. Vernon. "Heritability of the Big Five Personality Dimensions and Their Facets: A Twin Study." *Journal of Personality* 64 (1996): 577–591.

Jang, K. L., R. R. McCrae, A. Angleitner, R. Riemann, and W. J. Livesley. "Heritability of Facet-Level Traits in a Cross-Cultural Twin Sample: Support for a Hierarchical Model of Personality." *Journal of Personality and Social Psychology* 74 (1998): 1,556–1,565.

John, O. P., A. Angleitner, and F. Ostendorf. "The Lexical Approach to Personality: A Historical Review of Trait Taxonomic Research." *European Journal of Personality* 2 (1988): 171–203.

John, O. P., and R. W. Robins. "Gordon Allport: Father and Critic of the Five-Factor Model." In Craik, et al., *op. cit.* (215–236).

Johnson, J. A. "Ascertaining the Validity of Individual Protocols from Web-Based Personality Inventories." *Journal of Research in Personality* 39 (2005): 103–129.

Juntti, S. A., J. Tollkuhn, M. V. Wu, E. J. Fraser, T. Soderborg, S. Tan, S. Honda, N. Harada, and N. M. Shah. "The Androgen Receptor Governs the Execution, but Not Programming, of Male Sexual and Territorial Behaviors." *Neuron* 66 (2010): 260–272.

Kaffman, A., and M. J. Meaney. "Neurodevelopmental Sequelae of Postnatal Maternal Care in Rodents: Clinical and Research Implications of Molecular Insights." *Journal of Child Psychology and Psychiatry* 48 (2007): 224–244.

Kagan, J. *Galen's Prophecy: Temperament in Human Nature*. New York: Basic Books, 1994.

Kahneman, D. *Thinking Fast and Slow*. New York: Farrar, Straus and Giroux, 2011.

Kaminsky, Z. A., T. Tang, S. C. Wang, C. Ptak, G. H. Oh, A. H. Wong, L. A. Feldcamp, C. Virtanene, J. Halfvarson, C. Tysk, A. F. McRae, P. M. Visscher, G. W. Montgomery, I. I. Gottesman, N. G. Martin, and A. Petronis. "DNA Methylation Profiles in Monozygotic and Dizygotic Twins." *Nature Genetics* 41 (2009): 240–245.

Kammrath, L. K., R. Mendoza-Denton, and W. Mischel. "Incorporating If ... Then ... Personality Signatures in Person Perception: Beyond the Person–Situation Dichotomy." *Journal of Personality and Social Psychology* 88 (2005): 605–618.

Keirsey, D. *Please Understand Me II: Temperament, Character, Intelligence*. Del Mar, CA: Prometheus Nemesis Book Co., 1998.

Kelly, K. *Oprah, a Biography*. New York: Crown, 2010.

Kendler, K. S., C. O. Gardner, P. Annas, M. C. Neale, L. J. Eaves, and P. Lichtenstein. "A Longitudinal Twin Study of Fears from Middle Childhood to Early Adulthood: Evidence for a Developmentally Dynamic Genome." *Archives of General Psychiatry* 65 (2008): 421–429.

Kendler, K. S., K. Jacobson, C. O. Gardner, N. Gillespie, S. A. Aggen, and C. A. Prescott. "Creating a Social World: A Developmental Twin Study of Peer-Group Deviance." *Archives of General Psychiatry* 64 (2007): 958–963.

Kendler, K. S., K. Jacobson, J. M. Myers, and L. J. Eaves. "A Genetically Informative Study of the Relationship Between Conduct Disorder and Peer Deviance in Males." *Psychological Medicine* 38 (2008): 1,001–1,011.

Kim-Cohen, J., A. Caspi, A. Taylor, B. Williams, R. Newcombe, I. W. Craig, and T. E. Moffitt. "MAOA, Maltreatments, and Gene–Environment Interaction Predicting Children's Mental Health: New Evidence and a Meta-analysis." *Molecular Psychiatry* 11 (2006): 903–913.

Klein, J. *The Natural: The Misunderstood Presidency of Bill Clinton*. New York: Doubleday, 2002.

Klein, J. "Starting Over: Can Obama Revive His Agenda?" *Time*. January 21, 2010.

Kluckhohn, C., and H. A. Murray. "Personality Formation: The Determinants." In *Personality in Nature, Society, and Culture*. Edited by C. Kluckhohn, H. A. Murray, and D. M. Schneider. New York: Alfred A. Knopf, 1953.

Kouzarides, T. "Chromatin Modifications and Their Function." *Cell* 128 (2007): 693–705.

Kreisman, J. J., and H. Strauss. *I Hate You—Don't Leave Me: Understanding the Borderline Personality*. New York: Avon Books, 1989.

Kukekova, A. V., L. N. Trut, K. Chase, D. V. Shepelva, A. V. Vladimirova, A. V. Kharlamova, I. N. Osinka, A. Stepika, S. Klebanov, H. N. Erb, and G. M. Acland. "Measurement of Segregating Behaviors in Experimental Silver Fox Pedigrees." *Behavioral Genetics* 38 (2008): 185–194.

Laland, K. N., J. Odling-Smee, and S. Myles. "How Culture Shaped the Human Genome: Bringing Genetics and the Human Sciences Together." *Nature Reviews Genetics* 11 (2010): 137–148.

Lamason, R. L., P. K. Manzoor-Ali, M. A. Mohideen, J. R. Mest, A. C. Wong, H. L. Norton, M. C. Aros, M. Jurynec, X. Mao, V. R. Humphreyville, J. E. Humbert, S. Sinha, J. L. Moore, P. Jagadeeswaran, W. Zhao, G. Ning, I. Mkalowska, P. M. McKeigue, D. O. O'Donnell, R. Kittles, E. J. Parra, N. J. Mangini, D. J. Grunwald, M. D. Shriver, V. A. Canfield, and K. C. Cheng. "SLC24A5, a Putative Cation Exchanger, Affects Pigmentation in Zebrafish and Humans." *Science* 310 (2005): 1,782–1,786.

LeDoux, J. *The Emotional Brain*. New York: Simon & Schuster, 1998.

Lenneberg, E. H. *Biological Foundations of Language*. New York: Wiley, 1967.

Lesch, K. P., D. Bengel, A. Heils, S. Z. Sbol, B. D. Greenberg, S. Petri, J. Benjamin, C. R. Muller, D. H. Hamer, and D. L. Murphy. "Association of Anxiety-Related Traits with a Polymorphism in the Serotonin Transporter Protein Regulatory Region." *Science* 274 (1996): 1,527–1,531.

Lindberg, J., S. Bjornerfeldt, P. Saetre, K. Svartberg, B. Seehuus, M. Bakken, C. Vila, and E. Jazin. "Selection for Tameness Has Changed Brain Gene Expression in Silver Foxes." *Current Biology* 15 (2005): R915–R916.

Livesley, W. J. "Commentary on Reconceptualizing Personality Categories Using Trait Dimensions." *Journal of Personality* 69 (2001): 277–286.

Livesley, W. J. "A Framework for Integrating Dimensional and Categorical Classifications of Personality Disorder." *Journal of Personality Disorders* 21 (2007): 199–224.

Livesley, W. J., and K. L. Jang. "The Behavioral Genetics of Personality Disorder." *Annual Review of Clinical Psychology* 4 (2008): 247–274.

Livesley, W. J., K. L. Jang, and P. A. Vernon. "Phenotypic and Genetic Structure of Traits Delineating Personality Disorder." *Archives of General Psychiatry* 55 (1998): 941–958.

Livesley, W. J., K. L. Jang, D. N. Jackson, and P. A. Vernon. "Genetic and Environmental Contributions to Dimensions of Personality Disorder." *American Journal of Psychiatry* 150 (1993): 1826–1831.

Lykken, D. "A More Accurate Estimate of Heritability." *Twin Research and Human Genetics* 10 (2007): 168–173.

Lykken, D., M. McGue, A. Tellegen, and T. Bouchard. "Emergenesis: Genetic Traits That May Not Run in Families." *American Psychologist* 47 (1992): 1,565–1,577.

Lynam, D. R., and T. A. Widiger. "Using the Five-Factor Model to Represent the DSM-IV Personality Disorders: An Expert Consensus Approach." *Journal of Abnormal Psychology* 110 (2001): 401–412.

Maccoby, M. *The Productive Narcissist: The Promise and Peril of Visionary Leadership.* New York: Broadway Books, 2003.

MacDonald, K. "Evolution, the Five Factor Model, and Levels of Personality." *Journal of Personality* 63 (1995): 525–567.

McAdams, D. P. *The Stories We Live By: Personal Myths and the Making of the Self.* New York: Guilford Press, 1993.

McAdams, D. P., and B. D. Olson. "Personality Development: Continuity and Change over the Life Course." *Annual Review of Psychology* 61 (2009): 517–542.

McAdams, D. P., and J. L. Pals. "A New Big Five: Fundamental Principles for an Integrative Science of Personality." *American Psychologist* 61 (2006): 204–217.

McCrae, R. R., and P. T. Costa Jr. "Reinterpreting the Myers–Briggs Type Indicator from the Perspective of the Five-Factor Model of Personality." *Journal of Personality* 57 (1989): 17–40.

McCrae, R. R., and P. T. Costa Jr. "The Stability of Personality: Observations and Evaluations." *Current Directions in Psychological Science* 3 (1994): 173–175.

McCrae, R. R., and P. T. Costa Jr. "Personality Trait Structure as a Human Universal." *American Psychologist* 52 (1997): 509–516.

McCrae, R. R., and P. T. Costa Jr. *Personality in Adulthood: A Five-Factor Theory Perspective.* New York: Guilford Press, 2003.

McDonald, D. A., P. E. Anderson, C. I. Tsagarakis, and J. H. Holland. "Examination of the Relationship Between the Myers–Briggs Type Indicator and the NEO Personality Inventory." *Psychological Reports* 74 (1994): 339–344.

McGowan, P. O., A. Sasaki, A. C. D'Alessio, S. Dymov, B. Labonte, M. Szyf, G. Turecki, and M. J. Meaney. "Epigenetic Regulation of the Glucocorticoid Receptor in Human Brain Associates with Childhood Abuse." *Nature Neuroscience* 12 (2009): 342–348.

McGue, M., S. Bacon, and D. Lykken. "Personality Stability and Change in Early Adulthood: A Behavioral Genetic Analysis." *Developmental Psychology* 29 (1993): 96–109.

Maher, B. "The Case of the Missing Heritability." *Nature* 456 (2008): 18–21.

Mataix-Cols, D., L. Baer, S. L. Rauch, and M. A. Jenike. "Relation of Factor-Analyzed Symptom Dimensions of Obsessive-Compulsive Disorder to Personality Disorders." *Acta Psychiatrica Scandinavica* 102 (2001): 199–202.

Maynard Smith, J. *Evolution and the Theory of Games.* Cambridge: Cambridge University Press, 1982.

Mealey, L. "The Sociobiology of Sociopathy: An Integrated Evolutionary Model." *Behavioral and Brain Sciences* 18 (1995): 523–599.

Meaney, M. J. "Maternal Care, Gene Expression, and the Transmission of Individual Differences in Stress Reactivity Across Generations." *Annual Review of Neuroscience* 24 (2001): 1,161–1,192.

Meaney, M. J., and M. Szyf. "Maternal Care as a Model for Experience Dependent Chromatin Plasticity?" *Trends in Neuroscience* 28 (2005): 456–463.

Meyer-Lindenberg, A. "Genes and the Anxious Brain." *Nature* 466 (2010): 827–828.

Meyer–Lindenberg, A., J. W. Buckholtz, B. Kolachana, A. Hairi, L. Pezawas, B. G. Wabnitz, H. R. Verchinski, J. H. Callicott, M. Egan, V. Mattay, and D. R. Weinberger. "Neural Mechanisms of Genetic Risk for Impulsivity and Violence in Humans." *Proceedings of the National Academy of Sciences* 103 (2006): 6,269–6,274.

Miguel, E. C., J. F. Leckman, and S. Rauch. "Obsessive-Compulsive Disorder Phenotypes: Implications for Genetic Studies." *Molecular Psychiatry* 10 (2005): 258–275.

Miles, D. R., and G. Carey. "Genetic and Environmental Architecture of Human Aggression." *Journal of Personality and Social Psychology* 72 (1997): 207–217.

Miller, J. D., B. J. Hoffman, W. K. Campbell, and P. A. Pilkonis. "An Examination of the Factor Structure of Narcissistic Personality Disorders Criteria: One or Two Factors?" *Comprehensive Psychiatry* 49 (2008): 141–145.

Millon, T. *Personality Disorders in Modern Life,* 2nd ed. New York: John Wiley and Sons, 2004.

Millon, T., E. Simonsen, and M. Birket-Smith. "Historical Conceptions of Psychopathy in the United States and Europe." In *Psychopathy: Antisocial, Criminal and Violent Behavior.* Edited by T. Millon et al. New York: Guilford Press, 2002.

Mischel, W. "Toward an Integrative Science of the Person." *Annual Review of Psychology* 55 (2004): 1–22.

Mischel, W., and Y. Shoda. "Reconciling Processing Dynamics and Personality Dispositions." *Annual Review of Psychology* 49 (1998): 229–258.

Mischel, W., Y. Shoda, and P. K. Peake. "The Nature of Adolescent Competencies Predicted by Preschool Delay of Gratification." *Journal of Personality and Social Psychology* 54 (1988): 687–696.

Morf, C. C. "Personality Reflected in a Coherent Idiosyncratic Interplay of Intra- and Interpersonal Self-Regulatory Processes." *Journal of Personality* 74 (2006): 1,527–1,556.

Moffitt, T. E. "The New Look of Behavioral Genetics in Developmental Psychopathology: Gene–Environment Interplay in Antisocial Behaviors." *Psychological Bulletin* 131 (2005) 533–554.

Morris, J. A., C. L. Jordan, and S. M. Breedlove. "Sexual Differentiation of the Vertebrate Nervous System." *Nature Neuroscience* 7 (2004): 1,034–1,039.

Morris, J. A., C. L. Jordan, and S. M. Breedlove. "Sexual Dimorphism in Neuronal Number of the Posterodorsal Medial Amygdala Is Independent of Circulating Androgens and Regional Volume in Adult Rats." *The Journal of Comparative Neurology* 506 (2008): 851–859.

Munafo, M. R., S. M. Brown, and A. R. Hariri. "Serotonin Transporter (5-HTTLPR) Genotype and Amygdala Activation: A Meta-analysis." *Biological Psychiatry* 63 (2008): 852–857.

Munafo, M. R., B. Yalcin, S. A. Willis-Owens, and J. Flint. "Association of the Dopamine D4 Receptor (DRD4) Gene and Approach-Related Personality Traits: Meta-analysis and New Data." *Biological Psychiatry* 63 (2008): 197–206.

Mundy, L. "A Series of Fortunate Events." *Washington Post.* August 12, 2007.

Myers, I. B. *Gifts Differing: Understanding Personality Types.* Palo Alto, CA: Davies-Black Publishing, 1980.

Nader, R. *Crashing the Party: Taking on the Corporate Government in an Age of Surrender.* New York: St. Martin's Griffin, 2002.

Nettle, D. "An Evolutionary Approach to the Extraversion Continuum." *Evolution and Human Behavior* 26 (2005): 363–373.

Nettle, D. "The Evolution of Personality Variation in Humans and Other Animals." *American Psychologist* 61 (2006): 622–630.

Nicholson, I. A. M. *Inventing Personality: Gordon Allport and the Science of Selfhood.* Washington, D.C.: American Psychological Association, 2003.

Obama, B. *Dreams from My Father: A Story of Race and Inheritance.* New York: Three Rivers Press, 1995.

Obama, B. *The Audacity of Hope: Thoughts on Reclaiming the American Dream.* New York: Crown, 2006.

Oldham, J. M., and L. B. Morris. *New Personality Self-Portrait: Why You Think, Work, Love, and Act the Way You Do.* New York: Bantam Books, 1995.

Oler, J. A., A. S. Fox, and S. E. Shelton. "Amygdalar and Hippocampal Substrates of Anxious Temperament Differ in Their Heritability." *Nature* 466 (2010): 864–868.

Penke, L., J. J. A. Denissen, and G. F. Miller. "The Evolutionary Genetics of Personality." *European Journal of Personality* 21 (2007): 549–587.

Peper, J. S., R. M. Brouwer, D. I. Boomsma, R. S. Kahn, and H. E. Hulshoff Pol. "Genetic Influence on Human Brain Structure: A Review of Brain Imaging Studies in Twins." *Human Brain Mapping* 28 (2007): 464–473.

Peper, J. S., H. G. Schnack, R. M. Brouwer, G. C. Van Baal, E. Pjetri, E. Szekeley, M. van Leeuwen, S. M. van den Berg, D. L. Collins, A. C. Evans, D. I. Boomsma, R. S. Kahn, and H. E. Hulshoff Pol. "Heritability of Regional and Global Brain Structure at the Onset of Puberty: A Magnetic Resonance Imaging Study of 9-Year-Old Twin Pairs." *Human Brain Mapping* 30 (2009): 2,184–2,196.

Perani, D., and J. Abutalebi. "The Neural Basis of First and Second Language Processing." *Current Opinion in Neurobiology* 15 (2005): 202–206.

Peterson, C., and M. E. P. Seligman. *Character Strengths and Virtues: A Handbook and Classification.* New York: Oxford University Press, 2004.

Petrill, S. A., P. A. Lipton, J. K. Hewitt, R. Plomin, S. S. Cherny, R. Corley, and J. C. DeFries. "Genetic and Environmental Contributions to General Cognitive Ability Through the First 16 Years of Life." *Developmental Psychology* 40 (2004): 805–812.

Pincus, A., and E. B. Ansell. "Interpersonal Theory of Personality." In *Comprehensive Handbook of Psychology: Personality and Social Psychology,* vol. 5. Edited by T. Millon and M. Lerner. New York: Wiley, 2003.

Pinker, S. *The Blank Slate: The Modern Denial of Human Nature.* New York: Viking, 2002.

Plomin R., K. Asbury, and J. Dunn. "Why Are Children in the Same Family So Different? Nonshared Environment a Decade Later." *Canadian Journal of Psychiatry* 46 (2001): 225–233.

Plomin R., and D. Daniels. "Why Are Children in the Same Family So Different from Each Other?" *Behavioral and Brain Sciences* 10 (1987): 1–16.

Plomin, R., J. C. DeFries, G. E. McClearn, and P. McGuffin. *Behavioral Genetics,* 5th ed. New York: Worth Publishers, 2008.

Plomin, R. D., W. Fulker, R. Corley, and J. C. DeFries. "Nature, Nurture and Cognitive Development from 1 to 16 Years: A Parent–Offspring Adoption Study." *Psychological Science* 8 (1997): 442–447.

Preston, S. D., and F. B. M. de Waal. "Empathy: Its Ultimate and Proximate Bases." *Behavioral and Brain Sciences* 25 (2002): 1–72.

Renthal, W., I. Maze, V. Krishnan, H. E. Covington, G. Xiao, A. Kumanr, S. J. Russo, A. Graham, N. Tsankova, T. E. Kippin, K. A. Kerstetter, R. L. Neve, S. J. Haggarty, T. A. McKinsey, R. Bassel-Duby, E. N. Olson, and E. J. Nestler. "Histone Deacetylase 5 Epigenetically Controls Behavioral Adaptations to Chronic Emotional Stimuli." *Neuron* 56 (2007): 517–529.

Reynolds, S. K., and L. A. Clark. "Predicting Dimensions of Personality Disorder from Domains and Facets of the Five Factor Model." *Journal of Personality* 69 (2001): 199–222.

Rhee, S. H., and I. D. Waldman. "Genetic and Environmental Influences on Antisocial Behavior: A Meta-analysis of Twin and Adoption Studies." *Psychological Bulletin* 128 (2002): 490–529.

Ridley, M. *The Origins of Virtue: Human Instincts and the Evolution of Cooperation.* New York: Penguin, 1996.

Ridley, M. *Nature via Nurture: Genes, Experience, and What Makes Us Human.* New York: Harper Collins, 2003.

Riemann, R., A. Angleitner, and J. Strelau. "Genetic and Environmental Influences on Personality: A Study of Twins Reared Together Using the Self-Report and Peer-Report NEO-FFI Scales." *Journal of Personality* 65 (1997): 449–476.

Rinne, T., R. deKloet, L. Wouters, J. G. Goekoop, R. H. de Rijk, and W. van den Brink. "Hyperresponsiveness of Hypothalamic–Pituitary–Adrenal Axis to Combined Dexamethasone/Corticotropin-Releasing Hormone Challenge in Female Borderline Personality Disorder Subjects with a History of Sustained Childhood Abuse." *Biological Psychiatry* 52 (2002): 1,102–1,112.

Roberts, B. W., and A. Caspi. "The Cumulative Continuity Model of Personality Development: Striking a Balance Between Continuity and Change." In *Understanding Human Development: Life Span Psychology in Exchange with Other Disciplines.* Edited by U. Staudinger and U. Lindenberger. Dordrecht, The Netherlands: Kluwer Academic Publishers, 2003.

Roberts, B. W. and W. F. DelVecchio. "The Rank Order Consistency of Personality Traits from Childhood to Old Age: A Quantitative Review of Longitudinal Studies." *Psychological Bulletin* 126 (2000): 3–25.

Roberts, B. W., N. R. Kuncel, R. Shiner, A. Caspi, and L. R. Goldberg. "The Power of Personality: The Comparative Validity of Personality Traits, Socioeconomic Status, and Cognitive Ability for Predicting Important Life Outcomes." *Perspectives on Psychological Science* 2 (2007): 313–345.

Roberts, B. W., K. E. Walton, and W. Viechtbauer. "Patterns of Mean-Level Change in Personality Traits Across the Life Course: A Meta-analysis of Longitudinal Studies." *Psychological Bulletin* 132 (2006): 1–25.

Roberts, B. W., and R. W. Robins. "Broad Dispositions, Broad Aspirations: The Intersection of Personality Traits and Major Life Goals." *Personality and Social Psychology Bulletin* 26 (2000): 1,284–1,296.

Romeo, R. D. "Puberty: A Period of Both Organizational and Activational Effects of Steroid Hormones on Neurobehavioural Development." *Journal of Neuroendocrinology* 15 (2003): 1,185–1,192.

Sabol, S. Z., S. Hu, and D. H. Hamer. "A Functional Polymorphism in the Monoamine Oxidase A Gene Promoter." *Human Genetics* 103 (1998): 273–279.

Salvatore, G., G. Dimaggio, and A. Semerari. "A Model of Narrative Development: Implications for Understanding Psychopathology and Guiding Therapy." *Psychology and Psychotherapy: Theory, Research and Practice* 77 (2004): 231–254.

Samuels, J., G. Nestadt, O. J. Bienvenu, P. T. Costa Jr., M. A. Riddle, K. Y. Liang, R. Hoehn-Saric, B. Cullen, M. A. Grados, and B. A. Cullen. "Personality Disorders and Normal Personality Dimensions in Obsessive-Compulsive Disorder." *British Journal of Psychiatry* 177 (2000): 457–462.

Schmidt, L. A., N. A. Fox, K. Perez-Edgar, and D. H. Hamer. "Linking Gene, Brain, and Behavior: DRD4, Frontal Asymmetry, and Temperament." *Psychological Science* 20 (2009): 831–837.

Schmitt, D. P., J. Allik, R. R. McCrae, et al. "The Geographic Distribution of Big Five Personality Traits: Patterns and Profiles of Human Self-Description Across 56 Nations." *Journal of Cross-Cultural Psychology* 38 (2007): 173–212.

Schmitt, D. P., A. Realo, M. Voracek, and J. Allik. "Why Can't a Man Be More Like a Woman? Sex Differences in Big Five Personality Traits Across 55 Cultures." *Journal of Personality and Social Psychology* 94 (2008): 168–182.

Schmitt, J. E., L. T. Eyler, J. N. Giedd, W. S. Kremen, K. S. Kendler, and M. C. Neale. "Review of Twin and Family Studies on Neuroanatomic Phenotypes and Typical Neurodevelopment." *Twin Research and Human Genetics* 10 (2007): 683–694.

Schwartz, C. E., C. L. Wright, L. M. Shin, J. Kagan, and S. L. Rauch. "Inhibited and Uninhibited Infants 'Grown Up': Adult Amygdalar Response to Novelty." *Science* 300 (2003): 1,952–1,953.

Sen, S., M. Burmeister, and D. Ghosh. "Meta-analysis of the Association Between a Serotonin Transporter Promoter Polymorphism (5-HTTLPR) and Anxiety-Related Phenotypes." *American Journal of Medical Genetics,* Part B 127B (2004): 85–89.

Shaw, P., D. Greenstein, J. Lerch, L. Clasen, N. Gotgay, A. Evans, J. Rapoport, and J. Giedd. "Intellectual Ability and Cortical Development in Children and Adolescents." *Nature* 440 (2006): 676–679.

Shaw, P., N. J. Kabani, J. P. Lerch, K. Eckstrand, R. Lenroot, N. Gotgay, D. Greenstein, L. Clasen, A. Evans, J. L. Rapoport, J. N. Giedd, and S. P. Wise. "Neurodevelopmental Trajectories of the Human Cerebral Cortex." *The Journal of Neuroscience* 28 (2008): 3,586–3,594.

Shawn, A. *Wish I Could Be There: Notes for a Phobic Life.* New York: Viking, 2007.

Shifman, S., A. Bhomra, S. Smiley, N. R. Wray, M. R. James, N. G. Martin, J. M. Hettema, S. S. An, M. C. Neale, E. J. van den Ord, K. S. Kendler, X. Chen, and D. I. Boomsma. "A Whole Genome Association Study of Neuroticism Using DNA Pooling." *Molecular Psychiatry* 13 (2008): 302–312.

Shiner, R. L. "Linking Childhood Personality with Adaptation: Evidence for Continuity and Change Across Time in Late Adolescence." *Journal of Personality and Social Psychology* 78 (2000): 310–325.

Shiner, R. L. "A Developmental Perspective on Personality Disorders: Lessons from Research on Normal Personality Development in Childhood and Adolescence." *Journal of Personality Disorders* 19 (2005): 202–210.

Shiner, R. L., A. S. Masten, and J. M. Roberts. "Childhood Personality Foreshadows Adult Personality and Life Outcomes Two Decades Later." *Journal of Personality* 71 (2003): 1,145–1,170.

Shiner, R. L., A. S. Masten, and A. Tellegen. "A Developmental Perspective on Personality in Emerging Adulthood: Childhood Antecedents and Concurrent Adaptation." *Journal of Personality and Social Psychology* 83 (2002): 1,165–1,177.

Shweder, R. A. "Are Moral Intuitions Self-Evident Truths?" *Criminal Justice Ethics* 13 (1994): 24–31.

Shweder, R. A., N. C. Much, M. Mahapatra, and L. Park. "The 'Big Three' of Morality (Autonomy, Community, and Divinity) and the 'Big Three' Explanations of Suffering as Well." In *Morality and Health*. Edited by A. M. Brandt and P. Rozin. New York: Routledge, 1997.

Sisk, C. L., and D. L. Foster. "The Neural Basis of Puberty and Adolescence." *Nature Neuroscience* 7 (2004): 1,040–1,047.

Sisk, C. L., and J. L. Zehr. "Pubertal Hormones Organize the Adolescent Brain and Behavior." *Frontiers in Neuroendocrinology* 26 (2005): 163–174.

Skodol, A. E., J. M. Oldham, D. S. Bender, I. R. Dyck, R. L. Stout, L. C. Morey, M. T. Shea, M. C. Zanarini, C. A. Sanislow, C. M. Grilo, T. H. McGlashan, and J. G. Gunderson. "Dimensional Representations of DSM-IV Personality Disorders: Relationships to Functional Impairment." *American Journal of Psychiatry* 162 (2005): 1,919–1,925.

Sowel, E. R., B. S. Peterson, P. M. Thompson, S. E. Welcome, A. L. Henkenius, and A. W. Toga. "Mapping Cortical Change Across the Human Life Span." *Nature Neuroscience* 6 (2003): 309–315.

Spitzer, R. L., M. B. First, J. Shedler, D. Westen, and A. E. Skodol. "Clinical Utility of Five Dimensional Systems for Personality Diagnosis: A 'Consumer Preference' Study." *The Journal of Nervous and Mental Disease* 196 (2008): 356–374.

Srivastava, S., O. P. John, S. D. Gosling, and J. Potter. "Development of Personality in Early and Middle Adulthood: Set Like Plaster or Persistent Change?" *Developmental Psychology* 84 (2003): 1,041–1,053.

Steinem, G. *Marilyn*. New York: Henry Holt, 1986.

Steinberg, L. "A Behavioral Scientist Looks at the Science of Adolescent Brain Development." *Brain and Cognition* 72 (2010): 160–164.

Stout, M. *The Sociopath Next Door: The Ruthless Versus the Rest of Us*. New York: Broadway Books, 2005.

Sturm, R. A. "Molecular Genetics of Human Pigmentation Diversity." *Human Molecular Genetics* 18 (2009): R9–R17.

Tangney, J. P., J. Stuewig, and D. J. Mashek. "Moral Emotions and Moral Behavior." *Annual Review of Psychology* 58 (2007): 345–372.

Taylor, J., W. G. Iacono, and M. McGue. "Evidence for a Genetic Etiology of Early-Onset Delinquency." *Journal of Abnormal Psychology* 109 (2000): 634–643.

Terraciano, A., S. Sanna, M. Uda, B. Delana, G. Usala, F. Busonero, A. Maschio, M. Scally, N. Patriciu, W. M. Chan, M. A. Distel, E. P. Slagboom, D. I. Boomsma, S. Villafuerte, E. Silwerska, M. Burmeister, N. Amin, A. C. Janssens, C. M. van Duijn, D. Schlessinger, G. R. Abecasis, and P. T. Costa Jr. "Genome-wide

Association Scan for Five Major Dimensions of Personality." *Molecular Psychiatry* 15 (2010): 647–656.

Thomas, A., S. Chess, H. Girch, M. Hertzig, and S. Korn. *Behavioral Individuality in Early Childhood.* New York: NYU Press, 1963.

Thompson, P. M., E. R. Sowell, N. Gogtay, J. N. Gied, C. N. Vidal, K. M. Hayashi, A. Leow, R. Nicoloson, J. L. Rapoport, and A. W. Toga. "Structural MRI and Brain Development." *International Review of Neurobiology* 67 (2005): 285–323.

Trivers, R. L. "The Evolution of Reciprocal Altruism." *Quarterly Review of Biology* 46 (1971): 35–57.

Trut, L. N. "Early Canid Domestication: The Farm-Fox Experiment." *American Scientist* 87 (1999): 160–169.

Turkheimer, E., and M. Waldron. "Nonshared Environment: A Theoretical, Methodological, and Quantitative Review." *Psychological Bulletin* 126 (2000): 78–108.

Turri, M. G., N. D. Henderson, J. C. DeFries, and J. Flint. "Quantitative Trait Locus Mapping in Laboratory Mice Derived from a Replicated Selection Experiment for Open-Field Activity." *Genetics* 158 (2001): 1,217–1,226.

Visscher, P. M., S. E. Medland, M. A. Ferreira, K. I. Morfley, G. Zhu, B. K. Cornes, G. W. Montgomery, and N. G. Martin. "Assumption-Free Estimation of Heritability from Genome-wide Identity-by-Descent Sharing Between Full Siblings." *PLoS Genetics* 2 (2006): e41, 316–325.

Wallace, G. L., J. E. Schmitt, R. Lenroot, E. Viding, S. Ordaz, M. A. Rosenthal, E. A. Molloy, L. S. Clasen, K. S. Kendler, M. C. Neale, and J. N. Giedd. "A Pediatric Twin Study of Brain Morphometry." *Journal of Child Psychology and Psychiatry* 47 (2006): 987–993.

Weaver, I. C. G., N. Cervoni, F. A. Champagne, A. C. D'Alessio, S. Sharma, J. R. Seckl, S. Dymov, M. Szyf, and M. J. Meaney. "Epigenetic Programming by Maternal Behavior." *Nature Neuroscience* 7 (2004): 847–854.

Weedon, M. N., G. Lettre, R. M. Freathy, C. M. Lindgren, B. F. Voight, J. R. Perry, K. S. Eliott, R. Hackett, C. Giudicci, B. Shileds, E. Zeggini, H. Lango, V. Lyssenko, N. J. Timpson, N. P. Burtt, N. W. Rayner, R. Saxena, K. Ardlie, J. H. Tobias, A. R. Ness, S. M. Ring, C. N. Palmer, A. D. Moris, L. Peltonen, V. Salomaa, Diabetes Genetics Initiative, Wellcome Trust Case Control Consortium, G. Davey Smith, L. C. Groop, A. T. Hattersley, M. I. McCarthy, J. N. Hirschhorn, and T. M. Frayling. " A Common Variant of HMGA2 Is Associated with Adulthood and Childhood Height in the General Population." *Nature Genetics* 39 (2007): 1,245–1,250.

Weston, D., J. Shedler, and R. Bradley. "A Prototype Approach to Personality Diagnosis." *American Journal of Psychiatry* 163 (2006): 838–848.

Widiger, T. A., and S. N. Mullins-Sweatt. "Five-Factor Model of Personality Disorder: A Proposal for DSM-V." *Annual Review of Clinical Psychology* 5 (2009): 197–220.

Widiger, T. A., and D. B. Samuel. "Diagnostic Categories or Dimensions? A Question for the Diagnostic and Statistical Manual of Mental Disorders—Fifth Edition." *Journal of Abnormal Psychology* 114 (2005): 494–504.

Widiger, T. A., and T. J. Trull. "Plate Tectonics in the Classification of Personality Disorder: Shifting to a Dimensional Model." *American Psychologist* 62 (2007): 71–83.

Willis-Owen, S. A. G., and J. Flint. "Identifying the Genetic Determinants of Emotionality in Humans; Insights from Rodents." *Neuroscience and Biobehavioral Reviews* 31 (2007): 115–124.

Wilson, T. D. *Strangers to Ourselves: Discovering the Adaptive Unconscious.* Cambridge, MA: Harvard University Press, 2002.

Wolf, M., S. van Doorn, O. Leimer, and J. Wessing. "Life-History Tradeoffs Favour the Evolution of Animal Personalities." *Nature* 447 (2007): 581–584.

Wright, R. *The Moral Animal: Evolutionary Psychology and Everyday Life.* New York: Pantheon, 1994.

Wu, M. V., D. S. Manoli, E. J. Fraser, J. K. Coats, J. Tolkuhn, S. Honda, N. Harada, and N. M. Shah. "Estrogen Masculinizes Neural Pathways and Sex-Specific Behaviors." *Cell* 139 (2009): 61–72.

Yamagata, S., A. Suzuki, J. Ando, Y. Ono, N. Kijima, K. Yoshimura, F. Ostendorf, A. Angleitner, R. Riemann, F. M. Spinath, W. J. Livesley, and K. L. Jang. "Is the Genetic Structure of Personality Universal? A Cross-Cultural Twin Study from North America, Europe, and Asia." *Journal of Personality and Social Psychology* 90 (2006): 987–998.

Yang, J., B. Beben, B. P. McEvoy, S. Gordon, A. K. Henders, D. R. Nyholt, A. C. Heath, N. G. Martin, G. W. Montgomery, M. E. Goddard, and P. M. Visscher. "Common SNPs Explain a Large Proportion of the Heritability for Human Height." *Nature Genetics* 42 (2010): 565–569.

Zhang, T., and M. J. Meaney. "Epigenetics and the Environmental Regulation of the Genome and Its Function." *Annual Review of Psychology* 61 (2010): 439–466.

Zinsser, W. (ed.). *Inventing the Truth: The Art and Craft of Memoir.* New York: Houghton Mifflin Harcourt, 1998.

Acknowledgments

Making Sense of People grew out of my lifelong curiosity about personality differences. Writing this book gave me the opportunity to organize my thoughts and find out what's new in the field.

In surveying the vast literature on personality, I benefited greatly from the writings of many scholars, a few of whom I'd like to single out. On personality traits: Paul Costa, Lewis Goldberg, Robert McCrae, and Walter Mischel. On troublesome personality patterns: Aaron T. Beck, Theodore Millon, John Livesley, John Oldham, and Thomas Widiger. On the genetics of personality: David Goldman, Ken Kendler, Robert Plomin, and Daniel Weinberger. On stability and change in personality: Avshalom Caspi and Brent Roberts. On morality and character: Robert Cloninger, Jonathan Haidt, Christopher Peterson, Martin Seligman, Richard Shweder, and Frans de Waal. On identity and life stories: Dan McAdams. Thank you all.

This book was made possible by longstanding support from the University of California, which has provided me with an exceptional intellectual environment for more than 40 years—first at its San Diego campus (UCSD) and, since 1986, at its San Francisco campus (UCSF). Throughout this period I also benefited greatly from research funding from the National Institutes of Health and from private foundations, especially The McKnight Foundation. I would also like to acknowledge the friendship and support of Jeanne and Sandy Robertson, who helped me establish the Center for Neurobiology and Psychiatry at UCSF, and of Shari and Garen Staglin, who assisted in its further development through grants from The Staglin Music Festival for Mental Health and The International Mental Health Research Organization (IMHRO).

I discussed this book with many colleagues and received valuable suggestions from Steve Hamilton, Adrienne Larkin, John Livesley, Liz Perle, and Steve Rosen. I am also grateful for personal help from other colleagues: Peter Carroll, Glenn Chertow, Kerry Cho, Arkady Gendelman, James Ostroff, Lisa Vail, Nancy Ascher, Steve Tomlanovich, and Ayman Naseri. And I thank Jody Williams for assistance with the bibliography.

Special thanks go to my agent, Lisa Queen, for her perseverance, wisdom, thoughtfulness, good humor, and for guiding me through the changing world of book publishing. I also thank Kirk Jensen, who edited the first edition, and the outstanding professionals at Pearson who assisted in its production, especially project editor Andy Beaster, copy editor Chrissy White, cover designer Chuti Prasertsith, and compositor Nonie Ratcliff. And I am grateful to Amy Neidlinger who commissioned and oversaw the second edition.

Finally I'd like to thank a few members of my family, who each keep teaching me a lot about personality. I am fortunate to have my daughters, Elizabeth and Jessica, who are so close to each other and so close to me. I am fortunate to have my loving grandchildren, Jonah, Ellen, and Asher, my son-in-law Benjamin, and my stepson Whitney. And I am particularly fortunate to have my wife and soul mate, Louann Brizendine, who finishes my sentences, reads my mind, and keeps me continually inspired and entertained. My cup runneth over.

About the Author

Samuel Barondes is the Jeanne and Sanford Robertson Professor and Director of the Center for Neurobiology and Psychiatry at the University of California School of Medicine in San Francisco. He was trained in psychiatry and neuroscience at Columbia, Harvard, and the National Institutes of Health and has been at the University of California since 1970. He is the author of more than 200 research articles and has held many administrative and advisory positions, including Director of UCSF's Langley Porter Psychiatric Institute, President of the McKnight Endowment Fund for Neuroscience, and Chair of the Board of Scientific Counselors of the National Institute of Mental Health. He has received many honors, including membership in the National Academy of Medicine and the American Academy of Arts and Sciences. In addition to his research publications, Barondes has written three books about psychiatry for a general audience as well as a children's poetry book: *Before I Sleep: Poems for Children Who Think*. He lives in Sausalito, California, with his wife, Louann Brizendine.

INDEX